The Battle Rifle

ALSO BY RUSSELL C. TILSTRA

Small Arms for Urban Combat
*A Review of Modern Handguns, Submachine Guns, Personal
Defense Weapons, Carbines, Assault Rifles, Sniper Rifles,
Anti-Materiel Rifles, Machine Guns, Combat Shotguns,
Grenade Launchers and Other Weapons Systems* (McFarland, 2012)

The Battle Rifle

*Development and Use
Since World War II*

Russell C. Tilstra

McFarland & Company, Inc., Publishers
Jefferson, North Carolina

All photographs courtesy www.defenseimagery.mil

LIBRARY OF CONGRESS CATALOGUING-IN-PUBLICATION DATA

Tilstra, Russell C., 1968–
    The battle rifle : development and use since World War II / Russell C. Tilstra.
    p. cm.
    Includes bibliographical references and index.

    ISBN 978-0-7864-7321-2 (softcover : acid free paper) ∞
    ISBN 978-1-4766-1564-6 (ebook)

    1. Rifles—History—20th century. 2. Shooting, Military—History—20th century. 3. Rifles—History—21st century. 4. Shooting, Military—History—21st century. I. Title.

UD390.T55 2014
623.4'425—dc23
                              2014005652

BRITISH LIBRARY CATALOGUING DATA ARE AVAILABLE

© 2014 Russell C. Tilstra. All rights reserved

No part of this book may be reproduced or transmitted in any form or by any means, electronic or mechanical, including photocopying or recording, or by any information storage and retrieval system, without permission in writing from the publisher.

On the cover: 7.62mm FN SCAR-H Mk 17 battle rifle (On Point Firearms)

Manufactured in the United States of America

McFarland & Company, Inc., Publishers
Box 611, Jefferson, North Carolina 28640
www.mcfarlandpub.com

To Lorena and our monkeys.
I know you had to sacrifice family time.

Table of Contents

Preface 1

1. Post–World War II Western Rifle Development 5
2. FN FAL 16
3. HK G3 38
4. U.S. M14 51
5. SIG SG542 61
6. Galil 7.62mm NATO 71
7. FN SCAR-H (Mk17) 79
8. HK 417 87
9. Galil ACE 95
10. Post–World War II Combat Cartridge Development and Performance 100

Conclusion 159
Chapter Notes 169
Bibliography 175
Index 179

Preface

The twentieth century will forever be remembered as a time of great technological advancement. This applies to warfare as much as anything else, if not more so due to the lasting effects that wars bring. Despite numerous advances in military technology, a key weapon of war is still the individual soldier and his rifle. The basic form of the military rifle advanced during the last century like many other technologies.

While the year 1900 began with many of the world's military forces still using single-shot rifles, the manual repeating rifle was, even then, the new standard. In addition, automatic weapons like the belt-fed Maxim machine gun were already in service and there were those who knew, even then, that self-loading rifles would soon become practical for combat troops.

Though World War I was largely fought by troops using manual bolt action rifles, some light automatic weapons did see limited use, such as the Madsen light machine gun and the Browning Automatic Rifle (BAR). To be sure, though these weapons were portable, they weren't truly "light," weighing more than sixteen pounds without ammunition. Improvements would have to be made. By the start of World War II, some self-loading rifles were already in service, the most famous of these being the U.S. M1 Garand.

It was during World War II that the general concept of what a service rifle should be began to take on two different forms. The first became the dominant one, that of the "assault" rifle, the first of which was the German StG 44. This rifle concept would soon lead to the development of the most common rifle the world has ever known, the infamous Soviet AK-47.

The second form dominated the West during the Cold War. This type of shoulder arm came to be known as the "battle" rifle. The two are quite different. The assault rifle was meant to provide accurate semi-automatic fire to typical combat ranges of roughly three hundred meters; the AK-47 is effective to ranges well beyond this. The assault rifle was also intended to be capable of providing effective automatic "suppressive" fire meant to keep the heads of enemy troops down, allowing friendly forces to advance. This meant that the assault rifle had to be controllable when firing in the automatic mode. This requirement, by necessity, limits the power of the assault rifle and, as a result, its effective range.

In comparison, the Western (primarily U.S.) concept of the battle rifle was having the best of both worlds. The battle rifle was meant to provide the level of power seen in the bolt action rifles of World War I and, by extension, the U.S. M1 Garands of World War II fame. At the same time, the new rifle was to be capable of automatic fire. Unfortunately, it is not possible to have the best of both worlds; the higher power cartridges used in these rifles generate far greater recoil forces as well. This is a law of physics that, unlike the laws of man, cannot be broken. The problem created here was that the insistence on keeping the "giggle switch" on the battle rifle didn't work well in practice. The combination of the rifle's relatively light weight and the power of its round created a rifle that was all but impossible to control when fired on automatic.

As a result of this characteristic, many standard battle rifles were relegated to firing in the semi-auto mode only. The British choice of battle rifle, the L1A1, didn't even have the option of full-auto fire, and the standard U.S. M14 was often issued with a lockout device to prevent full-auto fire. This was after the U.S. had pushed hard to make the full-auto option a standard fitting.

By the end of the Cold War, most of the world had begun to give up on the battle rifle concept and its "unnecessarily" powerful cartridge and gone over to some form of assault rifle, the U.S. being the first country to implement the change, having done so by the end of Vietnam with the adoption of the M16A1.

In spite of this changeover, the basic understanding of the need for a powerful long-range service rifle was sound thinking, but battle rifles aren't very effective as automatic weapons except under very unique and limited circumstances, which will be covered in this book. What the world was to realize, however, is that the need for a long-range rifle still exists in some combat environments, much like the one the U.S. faces in Afghanistan today. Battle rifles were designed from the start to provide lethal rifle fire at over one thousand yards, and it is this particular characteristic that keeps the battle rifle in demand today. While most of the combat in the world today takes place in urban environments, there are times when engagement ranges are beyond the capability of the assault rifle. This is where the battle rifle comes into its own.

This book will examine the battle rifle on several levels. The first chapter examines the battle rifle's development following the end of World War II and will cover the whys of its recent return to service. It is followed by chapters that examine the most popular individual models in some technical detail. The next chapter provides historical development, with emphasis on key points in the evolution of rifle technology from the very beginning, ending with the crucial events that led to the development of the modern battle rifle. The final chapter covers ammunition developments in recent years and how these developments relate to the battle rifle and its current use.

Some recently introduced models have been covered only briefly, as there is not much history behind them, while other models have not been covered, as they are primarily commercially developed and not currently in military service or are used in very small numbers (the British L129A1 is a prime example). Some readers will no doubt feel some of these models have been overlooked and that is understandable, but these rifles will no doubt have their day should they prove superior designs.

While this book will not come close to achieving the detailed evaluation of works dedicated entirely to individual models like Lee Emerson's work on the U.S. M14 or R. Blake Stevens' volumes covering the FAL, it will give readers a thorough grasp of the battle rifle as a concept, as well as a good understanding of why it has seen a recent return to service. In addition, the reader should walk away with a strong knowledge base of the various models that are in service throughout the world and the realization that the battle rifle has far from outlived its usefulness.

1

Post–World War II Western Rifle Development

When World War II ended in Europe, the victorious nations used the opportunity to collect as much data from Germany as possible. This data covered many areas of study, from their rocket program and aeronautical research to small arms. It became clear from the evaluation of this research that the Germans had leaped ahead of much of the world in several technological arenas. This was especially true in the area of small arms development. While a good deal of this knowledge resulted from Germany itself confiscating the research of the nations it had occupied, it did some truly innovative work of its own. The study of modern infantry combat engagement ranges had led to the development of the StG 44 assault rifle.[1] This weapon was optimized for combat ranges of 350 meters and under. In Europe, where much of the terrain was fairly well wooded, this was more than acceptable for most combat conditions. The assault rifle concept was closely studied by both the U.S.S.R. and some Western nations, with the U.S. largely ignoring the concept.

The British felt that a slightly longer effective range was needed and following the war created a "Small Arms Ideal Calibre Panel" to develop a new military cartridge to arm their military.[2] They wanted a cartridge that was effective to six hundred meters, but they also wanted less recoil than that delivered by the .303 round that had been in use for over fifty years. They also desired a lighter rifle to chamber the new cartridge, one that could be selectively fired. Dr. Richard Beeching was one of the key figures on this panel and his team determined that the optimal combat caliber was a .270 diameter bullet. At an earlier time, American ordnance people had run similar tests and had decided that the ideal caliber was a .276, very similar in diameter (the M1 Garand test caliber). The British opted to go with the American diameter, though the name would later be changed to .280.[3] This is roughly around the time the British learned that the American military had no intention of using any caliber under .30, despite the results of the ballistic studies. Given the position of the U.S. at the time, both financially and politically, the British would lose out in their goal to adopt an intermediate cartridge. However, in 1951 the U.K. was set to adopt the .280 round and a new "bullpup" rifle to chamber it. This rifle was known as the EM2. Perhaps not coincidentally, the much later L85 (SA80) Enfield bore a great deal of resemblance to this futuristic rifle. Despite the efforts of the British, it soon became clear that this weapon/cartridge combination was not going to enter service. Several years later, in 1954, the experimental U.S. T65 round was adopted by NATO as the 7.62×51mm, known commercially as the .308 Winchester. While this new round had power nearly equal to that of the .30-06 used by the U.S. for the previous fifty years, it also had similar weight and generated similar recoil. This meant that a controllable selective-fire rifle was not going to appear anytime soon within NATO member arsenals.

In 1952, a series of rifle tests were held at the Fort Benning army base in Georgia. Four rifles were involved. The first place winner was to become the FN FAL, which was originally designed as a smaller rifle meant to chamber intermediate rounds like the .280 British and the StG 44 caliber, the 7.92×33mm. This design had to be enlarged considerably in order to handle the T65 cartridge used in these tests. Despite some indication of rigging the tests and some opinions that the FAL actually outperformed it, the experimental T44 would eventually become the U.S. M14.[4] The history of these rifles will be covered in more detail in their respective chapters.

During the height of the war, while the Germans were fielding the StG 44, another less expensive design was in development at the Mauserwerke in Oberndorf. This weapon, later known as the StG 45, had not yet been perfected by the war's end. The technology for this rifle was among the closely studied data mentioned earlier. One of Mauser's key people, Ludwig Vorgrimmler, had been moved out of Germany to continue his work, first to France, then later to Spain, where he worked at the Centro de Estudios Técnicos de Materiales Especiales (CETME). It was here that the roller-locked, delayed-blowback mechanism of the StG 45 was perfected. Spain was ready to adopt the CETME designed rifle, but they needed help with mass production techniques. Here is where the former Mauser crew stepped in. Most of these men had done some of the original work in Germany and had banded together again under a different banner. The firm of Heckler &

Koch (HK) was founded in 1949 in, of all places, Oberndorf. The founders were former Mauser people, as were some of the manufacturing people and machinists. They worked with CETME to further develop the new rifle design. The Spanish had wanted an intermediate round for the rifle, but the Germans wanted the weapon to be compatible with NATO requirements, so a compromise was reached. The CETME version would chamber the 7.62mm standard cartridge, but in a reduced power load. This was to become the Spanish model 58. Shortly afterward, HK reworked some of the bolt angles to allow the rifle to safely fire the full-power standard 7.62mm loading. This version would become the West German G3. The original West German rifle was the FAL, known as the G1 to the West German army. The FAL was replaced by the HK because FN and West Germany were at loggerheads over licensing local production of the G1.[5] The G3 was formally adopted by West Germany in 1959.

With these three primary rifles, the FAL, M14, and G3, the West was ready for the Cold War and their AK-47 opponent; at least that was the common misconception.

Oddly enough, shortly after the British were forced to bow to U.S. pressure and give up their well-researched quest for a .280 intermediate rifle/cartridge combination, the U.S. itself began to push for an even smaller, lighter, and less powerful round than what the British had desired. The concept of a very high-velocity .22 caliber centerfire round was gaining a small following within U.S. military circles. In 1952, the Operations Research Office (ORO), a short-lived think tank, conducted a study of recent military engagements and found results similar to those that led the Germans to develop the assault rifle concept in the first place. The ORO found that most firefights took place at under three hundred yards.[6] It can be assumed that the Korean War was the primary source for much of this data. This data was used in a 1959 release called "Operational Requirements for an Infantry Hand Weapon." The ORO had determined that a dispersion of projectiles would be more effective for making hits within 300 yards. This led to a project known as SALVO, which was to test the small caliber concept in more depth.

In the early 1950s, the U.S. Army had already shown interest in the same ideas regarding hit potential. Army Ordnance had tests conducted along this line of thinking. Researchers at the Aberdeen Proving Grounds test fired .22 caliber high-velocity rounds and studied the effectiveness of the bullets.[7] They liked the idea of their low weight and mild recoil. It was determined that these rounds were just as effective as larger calibers for "most combat ranges." The light weight of these rounds would allow the soldier to carry far more ammunition on patrols and combat operations. They would also allow for a smaller and lighter rifle from which to fire them, one that would be controllable in full-auto fire mode.

It was the results from these studies that ultimately led to the development of the M16. The M16's chief designer, Eugene Stoner from Armalite, had already developed a lightweight 7.62mm NATO rifle known as the AR-10, back in 1954. U.S. Army general Willard Wyman wondered if a smaller .22 caliber version could be built.[8] Stoner built one using a new .222 Remington Special round that would later become the .223 Remington. This was the birth of the 5.56mm cartridge used today. Stoner used the older .222 Remington as the basis for the new cartridge case design. He had to make it a bit larger to achieve the velocity that the Army wanted. Stoner ran into a great many roadblocks trying to get the new rifle accepted, mostly from Army general Maxwell Taylor, who was opposed to the rifle and its small cartridge.[9] The test weapon was known as the AR-15 and some military reports showed that the lightweight rifle was capable of delivering more firepower on target in a shorter period of time than the heavier M14/7.62mm combination. While this seems an unsound conclusion, the size and weight of the AR-15 and its ammunition was a major factor, as the AR-15 was quicker to put into action and, since more ammunition could be carried, more could be put on target. Colt Firearms had purchased the AR-15 production rights from Armalite's parent company Fairchild in late 1959[10] and needed to find customers soon, as the rifle looked set to die. However, Air Force general Curtis LeMay was a friend of Richard Boutelle, Fairchild's president. LeMay was at Boutelle's country estate celebrating Boutelle's birthday and the Fourth of July holiday, where he had a chance to see the AR-15 in action at one of Boutelle's shooting ranges on the property. He then quickly placed an order for thousands of the new design.[11] He wanted to replace the M1 and M2 carbines that were used for air base security at the time. These were getting old and LeMay probably just thought the AR-15 looked like an Air Force gun. It was sleek and modern and made of the same stuff as his airplanes. Wisely, President Kennedy turned down the request. The Air Force couldn't unilaterally adopt a new rifle, as the Army was the primary customer for such things. If the Army would request it, that was another story. General Taylor is generally accepted to be the man responsible for driving that issue home.

As all of this Washington game playing was unfolding, a more pressing issue was coming to the fore in Vietnam. After the French defeat at Dien Bien Phu in 1954, the U.S. began taking a more active role in Vietnamese affairs. We were already footing a good portion of the cost for continued French occupation. Increasing numbers of U.S. military advisors were being sent to Vietnam, and more and more were beginning to take an active role in field operations. Someone somewhere decided that the AR-15 would make a good jungle weapon. One of the reasons was that the light weight and low recoil of the AR-15 would be ideally suited to the small statured Vietnamese troops, as opposed to the M1 Garand, which weighed 9.5 pounds empty and kicked like a large-bore shotgun. It was more likely the active sales push from Colt and another involved company, Cooper-MacDonald, as both stood to profit from AR-15 sales.[12] Several samples were sent to Vietnam after recommendation by the Advanced Research Projects Agency (ARPA), which was set up to ensure that the U.S. was keeping up with technological advances around the world. One of ARPA's key early projects was known as AGILE. This project was intended to evaluate asymmetric warfare and offer advice on how to better fight such types of conflict. It was here that the AR-15 got its first chance to prove itself in the field. Reports came back that the AR-15 was perfect for the job of jungle warfare.[13] These reports are not historically known for their quality and accuracy.

In 1961, Defense Secretary McNamara and his "whiz kids" approved sending around one thousand AR-15s for issue to South Vietnamese troops.[14] More early reports came back praising the AR-15 and its combat qualities. Most important, it was claimed that the rifle was a great manstopper. The original rifling rate of one turn in fourteen inches of barrel was just enough to keep the bullet stable in flight. Once it connected with anything, it rapidly tumbled and broke apart. These characteristics, combined with a thin jacket metal, allowed for projectile fragmentation and an almost instantaneous transfer of energy leading to severe wounds. This earned the rifle the early nickname of "meat axe." Later models came with a faster one in twelve inch twist rate due to poor cold weather performance with the original rifling rate. This faster spin still did the job, however. It wasn't until the introduction of the new M855 62 grain bullet and the one in seven inch twist of the M16A2 and later models that the overstabilization of the projectile became a problem where stopping power was concerned. Even then, this was mostly applicable at ranges over two hundred meters. At that range, the bullet's velocity has slowed to the point that very little fragmentation occurs upon impact. The fragmentation issues for the 5.56mm bullets in general have never really been addressed, though many nations did not want the original M193 round or its notorious 55 grain bullet for "humane reasons."[15] Returning to the issue of the early AR-15 combat performance reports, project AGILE was highly supportive of the AR-15, to put it mildly. If taken at face value, one would have thought the new rifle was some kind of wonder gun. The two firms charged with selling the new rifle no doubt had a hand in the aggressive push concerning the capabilities of the new rifle and its cartridge. Robert MacDonald of Cooper-MacDonald was known to have personal contacts with several officers involved with project AGILE.[16] To be fair to the military, some officers were concerned with The Hague accord issues from early on. These voices were not the dominant ones at the time, however. No doubt some involved with the early projects just wanted to have their names thrown around the Defense Secretary's office. McNamara's office wanted to push for more tests comparing the AR-15 to the M14, which McNamara didn't care for, and the AK-47, which the U.S. intelligence and defense communities had foolishly ignored for years (in an Army report it was dismissed as a submachine gun).[17] The White House's interest in the AR-15 was no doubt due more to Kennedy's interest in counter-insurgency warfare and the "whiz kid" approach to solving issues.

So in the end, after the U.S. shoved the 7.62×51mm down NATO's throat despite objections, we wound up adopting a really unproven rifle that used a round even smaller than what the British had wanted in the first place. This must ultimately be blamed on the Kennedy administration and their game playing when this was not a game. Some Americans are no doubt still upset over such pathetic handling of the issue, especially given what took place after 1961.

Given his privileged background, the late President Kennedy was accustomed to snapping his fingers and having others jump. The late Secretary McNamara did not help the situation and, in fact, must be historically held responsible for canceling the M14's production and forcing the AR-15 (M16) on the U.S. soldier when it had clearly not yet been developed to the point of being service worthy. The whole thing was a rush job that resulted in the needless deaths of American servicemen. To make matters worse, once major problems with the M16 began surfacing, instead of pulling the rifles out of the field and re-issuing the M14 until the problems could be solved, the establishment tried its utmost to cover up the scandal. This was especially the case with one of the M16's early supporters, Army colonel Richard Hallock, since he was rightfully feeling somewhat responsible. He was not alone here, though. Army colonel Harold Yount was another key figure in the attempted cover-up. Hallock had been involved with project AGILE and helping push the AR-15 into service and Yount assisted Hallock in trying to keep the whole thing as quiet as possible. Yount was later removed from his position at Rock Island Arsenal.[18] It makes one wonder if General Westmoreland himself had been aware of the issues, as he was also a supporter of the M16, though there has not been any concrete evidence that he was directly involved with trying to cover up the M16 reliability problems. Sadly, the M16's failure issues were so numerous that the solutions were to be addressed without stopping the general issue of the rifle. This doesn't seem like a very wise decision from men who were supposed to be experts above all others. The most shameful episode in the M16 debacle was not Kennedy barking orders to his "whiz kids" to get something done without really knowing what he wanted done, nor was it McNamara forcing the unproven weapon system on the military simply because he didn't like the way the M14 rifle program had been handled, nor was it even the horribly unprofessional and conduct-unbecoming behavior of the officers involved with the attempted cover-up. Rather, the worst episode in the whole M16 tragedy was the attempt to blame the troops in the field for poor maintenance of the weapons and the reprimanding of U.S. Marine Corps officer Michael Chervenak, who was forced to go outside his chain of command to help bring the issue to light.[19] The first matter would have been rather hard to address, since there was very little in the way of proper cleaning equipment available at the time. The lack of M16 maintenance gear in the early years is well known by many historical accounts (it was promoted as self-cleaning). The attempt to blame the problems on improper maintenance was really just another part of the military's attempted cover-up anyway. Thankfully, the truth is impossible to conceal for very long. Troops in the field were still writing letters home to loved ones as they always do, and word still got around concerning the M16 failures. The officers attempting to cover up the matter should have been at least intelligent enough to know this much, but apparently, they weren't.

One of the many problems with the M16 was a combination of the dirty burning ball powder that had replaced the original extruded stick type powder of the early test ammunition. When this problem was combined with a lack of chrome plating in the bore and chamber area, the humid jungle climate, and the shortage of proper cleaning gear, the M16 didn't stand a chance; it was going to jam, and in short order.

One of the most common failures of this rifle at the time was a jam known as a failure to extract (FTE). This is when the empty, fired case remains in the chamber as the rifle's action is still functioning. This malfunction occurs when the extractor pulls free from its engagement with the case rim or, even worse, completely rips the case head off, leaving the body of the case inside the chamber (known as a case head separation). This last issue usually requires a tool known as a ruptured case extractor (aka stuck case puller) to remedy. This tool is usually only issued to the unit armorer and is not often seen in the field. Excessive pitting inside the chamber is usually the culprit in these malfunctions (chrome plating would have helped).

However, in the M16 design there is no primary extraction phase, which only serves to make matters worse. Lack of primary extraction is not unique to the M16. The AK series also lacks primary extraction. What happens without a primary phase is that the bolt simply unlocks and begins to rip the fired case from the chamber at very high speed. On early manual repeating designs like the model 98 Mauser, there is a primary phase when the extractor slowly pulls the case a small distance during the unlocking sequence. This helps to break the seal formed when the soft case metal has expanded to fit the chamber dimensions during the high-pressure, gas expansion phase of firing. By not having any primary extraction, the M16 is forced to operate by quickly and forcefully yanking the case from the chamber. If there is an excess of pitting in this area, enough said. The AK-47/AKM has a large extractor to help deal with this issue. The newer AK-74 series has an even more heavily built extractor, which was probably needed due to the steeper shoulder angle of the 5.45mm case when compared to the original 7.62×39 mm round. By 1968, many of the issues associated with the M16 had been addressed and the M16A1 (standardized in 1967) had begun to operate with a fair amount of reliability, as long as it was cleaned and lubed properly. The design remains sand sensitive to this day and still requires a great deal of lubricating oil in the proper places for reliable functioning. The oil must be applied regularly, as the heat builds up quickly due to design and burns the oil off rapidly.

As the war in Vietnam was coming to a close, the U.S., which had been so insistent about keeping a high-powered battle cartridge, had been the first NATO member to adopt a small caliber round. The rest of the member nations stayed with the 7.62mm until the 1980s. One of the concerns, as mentioned earlier, was the U.S. standard M193 cartridge and its 55 grain bullet that was known to lack desired armor penetrating ability. This round was also considered inhumane by some groups. Studies began on improving the 5.56mm in order to meet NATO requirements before it could be adopted as a second standard cartridge.

In 1980, the SS109 cartridge would see NATO adoption along with its 62 grain bullet complete with a steel penetrator tip inside. Unfortunately, this bullet required a very fast rifling rate for accurate shooting. One turn in seven inches was adopted as the ideal rate for this round, though a slower rate of 1-9" was used by some as an effective compromise. This slower rate was better for those nations that might be using stocks of the older M193 round, as accuracy was reduced when the lighter bullet was fired from the 1-7" barrel. The U.S. adopted the 1-7" rate for its M16A2, which became the new standard service rifle in 1982. This rifle also featured other changes, primarily in the areas of sights, furniture, and barrel dimensions. The new rifle also eliminated the full-auto capability, instead substituting a three-round burst option. While the burst feature was intended to limit ammunition consumption, it was not well received in some circles. It also had a poor side effect of lessening the quality of the trigger pull. Once the 5.56mm became NATO standard, several new rifle designs began to appear. Quite a few actually had their origins in the 1960s and '70s and some designs had already been finalized and adopted by this time. The HK 33 was available in the 1960s. The Italian Beretta AR70 and the Swiss SIG SG540 were available in the early '70s. The French FAMAS and the Austrian Steyr AUG were adopted in the late '70s.

While the West was working on once again following America's lead, the U.S.S.R. had also gone over to the small caliber, high-velocity concept with the adoption of the AK-74 and its notorious 5.45mm round. The Soviets had clearly been studying the M16 and its use in Vietnam. They astutely saw the advantage to the lightweight ammunition and the lighter recoil forces generated by the smaller rounds. They even seemed to have copied the terminal characteristics of the M193's bullet. The original AK-74 round (5N7) utilized a long bullet that was even lighter than the M193's. This bullet was known for causing horrible wounds and earned the nickname "poison bullet" in Afghanistan.[20] In the design of this bullet lies the key to its effectiveness. There was a small space in the interior nose and a small lead plug directly behind that. This was followed up with a mild steel core, and the whole thing was covered with a gilding metal jacket. When the bullet struck an object, the nose collapsed and the plug shifted and flattened, causing rapid yaw and consequent transfer of energy.

The original AK-74 rifle also spawned several variants. There was a folding stock version developed for paratroopers and special operations units known as the AKS-74. This model was fitted with a new pattern side folding stock as opposed to the bottom folding design seen on the earlier AKS-47/AKMS series. The design was considered stronger and was also used on the other early AK-74 variation, the AKS-74U (aka AKSU-74). This model is the one that was usually seen in the bin Laden home movies that he enjoyed sending to the media. Today's standard Russian Federation service rifle is the AK-74M, which is very much the same rifle as the original, though the side folding mechanism is now standard and the stock is of polymer construction, which more closely resembles a solid, fixed stock, and a left side receiver optic mounting rail has been added as well.

In Western matters, while the 5.56mm was considered to have more than enough range and power for a standard service cartridge, experience has shown otherwise. Up until the U.S. move into Afghanistan, most combat experience with the 5.56mm was within the typical modern combat ranges, meaning within three hundred meters, as the Germans had determined many years before. Even still, there were already complaints regarding the NATO standard SS109 cartridge and its U.S. counterpart, the M855. Some critics felt the bullet was overstabilized and had little stopping power beyond two to three hundred meters.[21] Once the U.S. became embroiled in combat in the open terrain of the Afghan mountains, the typical shooting ranges increased considerably (no one said the Afghanis were dumb). As a result, several new 5.56mm rounds have been issued on a limited basis. One of the more effective is the Mk262 with its 77 grain bullet. Engagement ranges are reported to be out to seven hundred meters for this round. While the Mk262 bullet is said to have slightly less armor piercing ability than the M855, its stopping ability is claimed to be better.

Despite such solutions, there has been a recent trend among U.S. troops and other allies involved as well. The switch back to 7.62mm NATO rifles has been gaining speed. In late 2009, the British army adopted a new rifle for issue to sharpshooters, the L129A1. This is an American design made by Lewis Machine & Tool (LMT). The initial order was for more than four hundred rifles. This weapon is based on the AR-10/AR-15 system and is very well made, with design emphasis on accuracy. The U.S. has been using the old M14 more now than at any time since the early years of the Vietnam War. There have been several versions of the M14 developed over the years and many of these have been pressed into service in Afghanistan.

Why this move between calibers yet again? The answer is simply that there is no such thing as a "do all" service rifle round. There never has been and, given current technology, there never will be. The 5.56mm was initially only required to perform out to five hundred yards (460m). During its baptism by fire in Vietnam, it is doubtful that the round was ever used at even that range. Most troops serving in Vietnam reported using the M16 at range of less than two hundred meters.[22] In Afghanistan, reports are that over half of all engagements begin at ranges over three hundred meters. This is the range at which the standard M855/SS109 round begins running into trouble when fired from a rifle length barrel. When fired from the 14.5" barrel of the M4/M4A1, the problems begin at ranges over 200m. This is why we are seeing a resurgent use of the 7.62mm NATO cartridge. There is no comparison between the small caliber rounds used in weapons like the M16 and AK-74 and the much more powerful .30 caliber cartridges used in such rifles as the M14, FAL and G3, and Russian SVD sniper rifle. While this last rifle is still standard, the previous three have been largely replaced or at least been moved to second line service with most nations. At the very least, they have seen an end to mass production, though some production does continue in various developing nations that had production facilities set up in earlier years. The parent nations of these designs have switched over to more modern 5.56mm models. This development and the need for a more powerful rifle in the Middle East has led to a new mini-wave of 7.62mm NATO designs to help fill the gap. Combat conditions in Afghanistan have shown the world that the 7.62mm NATO is far from finished as a service rifle caliber. The situation has also shown that while statistics may provide an indication of the norm (e.g., engagement ranges), there are always exceptions to the rule. With the continued U.S. involvement in Afghanistan, and with a possible Iranian-Israeli conflict on the horizon, the career of the 7.62mm NATO service rifle is far from over.

The remainder of this book will examine the most successful 7.62mm NATO rifles along with some new designs that may very well have a future as issue weapons. There will also be a short chapter dealing with some of the newer calibers developed that have attempted to provide the services with a solution to the range/lethality issues of the 5.56mm and the excessive recoil of the 7.62mm NATO rifle.

2

FN FAL

The Fabrique Nationale (FN) Fusil Automatique Léger (FAL) was the most popular Western rifle of the post–World War II period. The origins of this rifle can be traced to the FN model 49 (aka SAFN). Design of the model 49 began prior to the war. The model 49 was largely the work of FN engineer Dieudonne Saive, the man credited with completing the work begun by John Browning on the Hi-Power pistol. Production of the model 49 was put on hold due primarily to the war. After Germany's defeat, Saive completed his work on the model 49 and it was put into production, but while it was successful, it would soon be made antiquated by an even better design, the FAL.[1] Oddly, while they both used a tilting bolt action, the FAL was originally intended to use a smaller intermediate round in the true sense of an assault rifle, while the model 49 was designed to use much higher-powered cartridges. In fact, the original test model of the FAL used the German 7.92×33mm kurz (short) round first developed prior to World War II.[2] Only a few weapons were ever designed for this cartridge. The primary weapon for this round was the StG 44. Other German gun makers did put forth assault rifle designs to utilize this round, but its importance in the firearms industry was as a technological and tactical breakthrough, one that would greatly affect weapons design to follow, especially in the Soviet AK-47, the world's most widespread firearm.

The original intermediate caliber FAL was not to see a future, however, and the world became familiar with the larger, heavier production model. This rifle was to become the primary infantry weapon for much of the free world, for many years. As many FAL fans already know, close to one hundred nations would come to use this rifle in one form or another, with licensed production taking place in many countries. Among the nations that would manufacture the FAL were Argentina, Australia, Austria, Brazil, Canada,

The FN FAL in its basic fixed stock variation. This rifle served much of the free world for the greater part of the Cold War, rivaled only by the Heckler & Koch G3. The grooves running along the forend are for housing an optional folding bipod. While somewhat sand sensitive, the FAL brings with it a great deal of power, range and proven performance, all of which are desperately needed by combat troops serving in Afghanistan. (TSGT Ken Hammond)

Greece, India, the U.K., South Africa, and Venezuela.[3] With regards to the different versions of the FAL, it should be understood that there were two primary patterns of FAL: the metric pattern and those produced by Commonwealth nations. These last are often called "inch" pattern FALs. While there was very little difference dimensionally between the two, there were design differences, enough so that particular parts are often not interchangeable between the two types.[4] When these differences are combined with the seemingly endless model variations of the FAL, classifying these rifles becomes difficult and tedious at best. I will, however, attempt to go over the basic and most common variants and cover the primary differences between the most common military variations used over the years. Later I will cover the civilian variants that have been or are still available to the public.

After the initial FAL prototype, further developed models were made for the British military in their experimental .280 caliber. This was the same round as used in their EM2 bullpup. One of these FAL prototypes was also a bullpup in layout, while the second model was a conventional rifle.[5] For any readers unfamiliar with the term "bullpup," it refers to a rifle with its operating action located behind the trigger group, resulting in a short overall length while allowing for a full-length barrel. While the bullpup FAL did not see any success, the standard layout continued to see development. The British desire to use the EM2 may have been partially behind the demise of the FAL bullpup prototype. When the U.S. was given the FAL prototype for testing, it soon became clear that our military had no intention of using an intermediate round for standard issue. Given this, FN's designers went to work yet again and reworked the FAL to handle the T65 experimental round that would later become the 7.62 NATO. By this time the FAL had grown to over nine pounds empty weight and just under forty-three inches in overall length. It was no longer an assault rifle, it had become a battle rifle, and as a result interested nations like the U.K. would drop the full-auto position on the selector switch, as a conventional 7.62 NATO rifle is hardly effective on full-auto at all but the closest ranges. After initial testing in the U.S., there were some in our military who felt it was a superior weapon when compared to competing American designs. Due to this interest in the FAL, a royalty free production deal was reportedly offered by FN. Small production runs of the FAL were set up in the U.S. by Harrington and Richardson (H&R) and High Standard. These were named the T48, and they would eventually lose out to the T44E4, which would eventually become the M14 rifle in 1957, though there were some in the U.S. who felt this was a mistake, not only from a design standpoint but politically as well. By this time it had been assumed that the U.S. would agree to adopt the FAL if the other NATO countries would adopt the 7.62×51mm cartridge, which they did in 1953.[6] When the dust settled, the U.S. did not accept the FAL and NATO members were not to see a standard rifle-cartridge combination.

Though battle rifles in general are nearly useless when fired in full-auto mode, mention should be made of the FAL's cyclic rate, which is roughly 650 rpm at the low end. Its rate of fire can be considerably faster if the gas system is set to greater levels.

As many experienced troops already know, automatic fire has its very limited and specific place, and due to their combination of high power, heavy recoil and relatively light weight battle rifles have even more limited roles when firing in full-auto mode.

While a little off topic, it must be mentioned that automatic fire was used to a horribly effective level during World War I. This was largely due to the tactics in place at the time. Men charging over open terrain directly into a machine gun's line of fire stand very little chance of survival. It must be remembered, however, that the machine guns in use at the time, such as the Browning M1917, the British Vickers, and the German Maxim MG08, were extremely heavy and usually water cooled and mounted to an equally heavy tripod. The consequent weight and rigidity that resulted from this combination made for very controllable weapons. The relatively slow rate of fire (below 600 rpm) of these early automatic weapons also contributed to their controllability. The troops that went up against these weapons initially had little concept of what the guns could do to a massed infantry assault. Many had never encountered such weapons before, as machine guns had only been invented roughly thirty years earlier and had never been used on this scale before. They learned very quickly, though the American Indian could have offered some advice here, as they had prior experience with a similar weapon invented by Dr. Richard Gatling.

The battle rifle, however, when compared to these much heavier and more controllable, though slower firing, weapons, weighed in the neighborhood of nine pounds and fired 100–200 rpm faster, with recoil forces that were almost identical. Remember, the 7.62mm NATO is almost the ballistic equivalent of the caliber in which the Browning M1917 machine gun was chambered. Given this, it is relatively easy to see why a battle rifle's effectiveness when fired in full-auto is severely limited. Troops in Vietnam were filmed in combat firing the M14A1 (SAW variation) in full-auto bursts and the striking rounds could be clearly seen hitting the ground just yards in front of the shooter. This was nothing more than wasting ammunition. The soldier was likely new, but this was a lesson many learned when attempting to fire the M14 in automatic mode. It was nearly impossible to control.

The final verdict was that automatic fire from a battle rifle is most often highly inaccurate and is primarily useful as suppressive fire, and then usually only at point blank range. However, point blank range for a 7.62mm NATO rifle can be upward of fifty yards. Power is not something this cartridge lacks.

This lack of control in full-auto was one of the original motivating factors for the development of the assault rifle. Nazi Germany was well aware of the limitations of high-power rifles when attempts were made to fire automatically.

While a sixteen- to eighteen-pound Browning Auto Rifle (BAR) could reasonably handle the recoil of its .30-06 chambering, when such a powerful cartridge was fired in a rifle of conventional weight (eight to ten pounds) things were very different. The Germans did field the FG-42 (a full-power selective-fire rifle) in very limited numbers, primarily to its airborne units. The success with such weapons was mixed , and the concept never caught on after the war. The conclusion to this matter is that the FAL performed best in semi-auto mode, as do all battle rifles. While some nations adopted select-fire versions of the FAL, the British opted for a semi-auto only rifle when they chose the L1A1 as their standard service rifle.

During this time, the FAL was beginning to see widespread adoption, with Canada being the first large customer in 1954.[7] Many South American nations would soon follow as well. As political unrest was becoming common in many parts of the world at this time, the FAL was becoming a common sight, especially in the hands of soldiers in the U.K. as the L1A1 Self-Loading Rifle (SLR). The Australians would put the FAL to good use during Vietnam and many Aussie troops would come to appreciate the power of the 7.62 NATO round and its ability to penetrate barriers.

While the FAL rifle was a great success, its heavy barrel squad automatic weapon (SAW) variation did not achieve the same reputation. While a large number of these weapons were produced, they had a nasty reputation for jamming on the third round of a full magazine when fired on full-auto. This was often called the "bang-bang, jam" issue.[8] Reasons for this shortcoming were never fully solved. This chapter is meant to cover only the rifle version, but mention should be made of this flaw. This trait, combined with the added weight and lack of a quick change barrel, made this version of the FAL a poor choice for a SAW. As for the standard rifle model, it was a huge success with a total production estimate of over 2 million manufactured between 1953 and the present.[9]

As for the technical aspects of the rifle, it is a short-stroke gas-operated weapon available in select-fire or semi-automatic only variations. As mentioned, the selective-fire models were too light to be controllable when fired in the fully automatic mode. The earliest production FALs utilized wooden buttstocks, but plastic soon became standard. The pistol grips were also made of plastic, as were the forearms. Some nations, like Austria and the Netherlands, used sheet metal forearms as standard. The Israeli-made models used a combination of wood and metal. The Israeli variation was adopted in the 1950s and local production of many components began soon after.

The cocking handle on the FAL was located on the left side of the receiver to allow the user to keep hold of the pistol grip (for right-handed shooters). During firing, the cocking handle was stationary so as not to interfere with the shooters' aim and as an added safety measure. Many FALs came with a carrying handle that folded to the right side and was roughly located at the longitudinal center of gravity. Some users removed this feature to keep it from snagging or possibly making noise during patrols. On some Para versions no carrying handle was supplied, but one could be added if desired, as the recess cut in the receiver remained.

The standard FAL used a fairly long barrel of approximately 21 inches, though shorter models were offered. Handy 17"–18" carbine models were first produced in the early 1960s. The model 50-63 Para was the best known of these shorter rifles. The Para versions used a sturdy right-side folding stock. The folding models were also made with standard length barrels as well (50–61). There was a lightweight folding model known as the 50–64. This model had the standard length barrel with a lower receiver using a lightweight alloy in place of steel.[10] This presented no problems, as the lower receiver received very little stress from firing.

The use of a folding stock necessitated a change to the weapon's internal design. On the original fixed stock model, the recoil spring was housed in the buttstock itself. This could not be the case for the folding model, so the receiver and bolt carrier had to be modified to house a new recoil spring design. This new design proved equally dependable in service, which is often not the case when redesigning a weapon, and due to its success both designs are still in use today.

A brief rundown of the rifle's operating sequence is in order. First ensure that the weapon is on safe and insert a loaded magazine. The FAL's magazine must be inserted nose first and then rotated up into place. After you ensure that the magazine is locked in its well, the cocking handle is pulled fully to the rear and released, though the bolt catch can also be used to release the carrier if it was previously engaged, holding the carrier to the rear. The selector lever is then moved to either the semi-auto or auto position and then the rifle can be fired when the shooter is ready.

On the FAL, the bolt remains open after the last round is fired, with the exception of some models that lack this feature. There is no full-auto selector position on the British L1A1 and other similar models. The magazine release is a lever located directly behind the magazine, while the bolt release is a plunger type button located just left of the magazine release. Pulling down on this plunger releases the bolt catch. The bolt catch can also be manually engaged by holding back the bolt and pushing up on the catch, then allowing the bolt to ride forward.

As mentioned, the FAL is a short-stroke gas-operated rifle, which uses an adjustable gas regulator to ensure reliable functioning under varying conditions of fouling. There was also a "G" shut-off position on the gas plug used for grenade launching, with "A" for normal firing. As the FAL is fired, the bullet passes the gas port in the barrel and a small amount of gas enters the gas plug. The expanding gas then pushes the piston rearward in the usual fashion. The gas piston is connected to a long piston rod, which pushes the bolt carrier to the rear. A return spring surrounding the piston rod returns the piston to battery, while the carrier continues on unlocking the bolt by cam action, extracting and ejecting the empty cartridge case.

The gas system can be adjusted to allow for a larger or smaller volume of gas in operating the piston. Correct adjustment of the rifle's gas system is needed to minimize abuse of both the shooter and the rifle's working parts.

The well-known method for adjusting the gas system is to start with an empty magazine in the rifle and the gas system at its lowest setting with the regulator knob turned up against the front sight base. This knob is notched and spanner wrench is made for this purpose, but almost any small tool can perform this function, such as a cartridge. Single rounds are then fired with the shooter backing off the gas regulator knob one click at a time until the bolt catch is activated. This means the rounds must be manually loaded into the breech each time. After the correct setting is reached and the bolt catch is activated, one or two more clicks should be added to ensure reliable operation. Australian troops used to play a joke on American soldiers during the Vietnam War by turning up the gas regulator all the way and then letting the curious American grunt fire the rifle. The recoil was rather severe, to say the least. When the gas system is properly adjusted, the FAL is claimed to have more mild recoil than the HK G3, which has a reputation for having rather harsh firing characteristics, a consequence of its design. The FAL's balance and trigger are generally considered superior to the HK as well. It is often considered to be a better handling rifle than the M14, though the FAL's trigger and sights are not of the same caliber (pun intended) as those of the M14.

The bolt on the FAL locks against a fitted shoulder by a tilting motion with unlocking taking place as the bolt is cammed upward by ramps machined into the bolt carrier. The M14, by comparison, utilizes a rotating bolt design.

The trigger group is a relatively simple design utilizing a double sear setup with the auto (safety) sear holding the hammer back until the bolt is fully locked. The trigger sear is the other sear, which both pivots and slides. The sliding motion is needed to provide a disconnector function. As it slides forward, the sear pivots up to engage the sear notch on the hammer. As the front of the sear pivots up, the back drops down into a short step in the trigger proper. To allow for this, there are actually two lengths of trigger pull, one for full-auto and one for semi-auto, with the full-auto pull being a bit longer. The selector lever acts as the trigger stop.

While the FAL is fairly accurate, it generally ranks behind the M14 and the HK G3 in this regard. This is no doubt an arguable generalization. Consequently the FAL has not been terribly successful when attempts were made to turn it into a sniper rifle. There were two primary reasons for this reputation of mediocre accuracy. The first was due to attempts to mount a scope by using the thin sheet metal upper receiver cover as a mounting point. There was simply no rigidity when mounting a scope in this fashion and scopes would never hold a zero as a result. Later mounting variations helped in this regard by using a more rigid wraparound mount (Leatherwood mount).[11] Despite this fix, the FAL suffers from another, less severe accuracy issue. Due to the tilt locking bolt, as the magazine was emptied less pressure was placed on the bolt from beneath by the rounds remaining in the magazine. As a result, the bolt actually shifts slightly with each shot. This alters the pressure being placed on the round in the chamber. The variation from shot to shot isn't great, but it becomes more pronounced at longer ranges.

The FAL has used several types of iron sights, depending on the particular model. The most common were both the low- and high-profile standard sights that had an aperture that slid on a rail to allow for rapid adjustments in range. There was also a fixed elevation 300 meter sight, popularly known as the Holland sight. The Para model FALs usually used a two position rear sight with settings for 150 and 250 meters.[12] Adjustments in windage for FAL rear sights were accomplished by loosening and tightening opposing screws. Elevation zeroing was done through the front sight. The main flaw in the sliding model rear sight was the play built into the design. While the FAL's sights were decent, they were not as precise as those of the M14. They still did the job, however, and proved rugged enough for battle conditions.

The FAL has seen combat use since its first adoption in the mid–1950s. Almost without exception, it has given sterling service regarding its combat performance. Despite the FAL's poor arctic performance during U.S. testing in 1953, later production models appeared to have served quite well in cold climates. Again, these tests may have been skewed somewhat to favor the T44 (M14).[13] The FAL has shown good performance in jungle climates as well.

The one climate where the FAL suffered reliability issues was the desert. This is rather odd, as the FAL was produced by both Israel and South Africa and served as standard issue in both nations for much of the 1960s and '70s. Not surprisingly, it was eventually replaced by an AK derivative in both nations. In fact, it was replaced by the same rifle in both nations, the Israeli Galil. The South African produced version of the Galil was the R4, a slightly modified model. During the Six-Day War, many Israeli troops complained of the FAL jamming in sandy conditions. It is likely that the reason for these complaints may have been the intensity of the fighting, where there was little time to clean the weapon properly, if at all. Apparently using grooved bolt carriers, which allow for the accumulation of grit, did not improve things. These carrier types were already being used on British models. By this time, the Israelis had become so impressed with the performance of the AK-47 that it is no surprise their next service rifle was based on this design. The Galil and R4 are often considered the most durable and reliable military rifles ever made. Most of the rest of the world, however, continued to use either the AK-47, the FAL, or the HK G3. In all fairness to the FAL, desert conditions have caused many decent designs to stop cold. The large amount of clearance built into the AK's design is likely the primary reason for its continuing to reliably function in sandy environments. Only the most rugged designs have been able to reliably operate with any consistency in sunny desert "climes."

While the FAL has seen several modifications over the years to make it less sand sensitive, none of these changes were to its basic operating system, which remained unaltered throughout its production history. The changes were merely cuts in surfaces like the bolt carrier, to allow for the accumulation of sand and grit.

The issues regarding automatic weapons' reliability in any environment are complicated at best. The problems are compounded under severe climate conditions. This was the case with the U.S. M1 and M2 carbines on the Korean Peninsula in the early 1950s.

The M1/M2 carbine family had proven reliable enough during World War II but had developed a rather poor reputation for stopping power early on. This issue aside, the carbine acquired a reputation for jamming under conditions of extreme cold like that present during Korean winter conditions. The problem behind this was twofold and one of these issues is of particular importance as it is also a factor regarding the reliability of the FAL under desert conditions.

The two primary forms of gas operation for automatic rifles are short-stroke piston and long-stroke piston. Direct impingement like the system used on the M16 series is a separate type of gas operation that does not utilize a piston and operating rod; though some argue that the M16 is not a true direct impingement system, the end result is the same, with fouling and heat being forced into the receiver area during firing. At any rate, the two primary methods have remained and both continue to use a conventional piston and operating rod.

Both of these primary systems come in various forms, but the primary difference between the two is in the attachment of the piston to the bolt carrier parts group.

The most often given definition for a long-stroke gas system is a design having the piston attached to the bolt carrier. For a short stroke, the two parts are separate and the piston does not travel with the carrier during the recoil phase of operation.

By definition, both the FAL and the M1/M2 carbine are both short-stroke weapons, though their systems are considerably different in design.

The U.S. M1/M2 carbine uses a non-adjustable, captive gas tappet piston design, while the FAL gas piston design is fully adjustable and is connected to an operating rod. The M1/M2 has no such operating rod. With the M1/M2 family, the piston impacts a flat recessed portion of the "operating slide." This is the name given to the part that acts as a combination operating rod, bolt carrier, and cocking handle.

One of the determining factors that led to the U.S. M1/M2 carbine jamming during the Korean War had to do with weak operating springs. This is not a great surprise, as all carbines in service at the time had been manufactured prior to 1946 and were four years old at best. Some were made as early as 1941.

The other factor determined to be responsible for the M1/M2 carbine's reliability issues had to do with the characteristics of operating gases in low temperatures. Under very cold conditions, gas expands more slowly and, as a result, far less force is available for delivery to the operating parts. This characteristic is of great concern, as it will affect any gas-operated firearm, including the FAL.

This factor, however, affects short-stroke systems more severely than it does long-stroke systems. This is regardless of their respective designs. As the short-stroke systems generally have less mass moving during much of the recoil phase, recoil spring force becomes more crucial. In a short-stroke system, if less initial force is being supplied to the operating parts jamming is more likely to occur. This is more of a problem for the U.S. M1/M2 carbine, as it did not have any adjustment capability with regard to its gas system. The FAL, on the other hand, does allow for a greater volume of gas to be delivered to the gas piston, in effect helping to overcome the cold temperature behavior of the expanding gases needed for reliable operation of any gas system.

The effects of this cold weather characteristic aren't seen as drastically on long-stroke gas-operated designs due to the increased mass of the reciprocating parts. Another crucial factor in the reliable functioning of any gas-operated design is the proper alignment of the gas system in relation to the receiver. If the rifle is dealt a severe blow in the right (or wrong) location, the alignment of the gas system can be offset and reliability will suffer due to increased friction between operating components during the cycling of the action. This increased friction can also be caused by the sludge built up from the mix of lubricating oil and sand (remember the Israeli FAL). It is this factor that causes the majority of issues regarding the reliability of self-loading weapons under desert conditions.

This is why grooves have long been cut into receivers and bolt carriers. These grooves act as a trap for the inevitable buildup of sludge when the oil and sand mix together. Such grooves are not only present on some FAL bolt carriers, but similar grooves were designed for the Israeli UZI submachine gun from the start. These grooves are formed into the UZI's receiver when it is initially stamped. A similar process takes place when forming the receiver of the Italian Beretta M12S, another popular submachine gun design known for its ability to accept large amounts of fouling without jamming (the M12S's issue with the ejector jams is another story). The receivers on these two particular weapons are stamped, and as stamped parts they do not require the grooves to be cut as a separate machining process. This saves both time and money. On the FAL, the grooves must be milled at a later time. The presence of these "grit traps" no doubt helps in the weapon's ability to continue to operate reliably under adverse conditions.

While the intensity of the fighting during the Six-Day War may have helped expose a weakness in the FAL's design (sand sensitivity), this is not definitive and likely not the sole reason for the problems experienced by the Israeli troops during this war. Many other factors are involved in this equation. As mentioned, this fighting was prolonged and intense for the Israelis. Proper maintenance was most likely non-existent for the FALs during this short war. The troops didn't dare risk taking their weapons apart for proper cleaning, for an attack could occur at a moment's notice. This alone would have affected the functioning of most weapons, though the AKs and UZIs in use during this conflict appear to have fared better in this regard.

Also, troops involved in the fighting no doubt had to rush through any basic maintenance procedures being performed. They likely feared any field stripping of the FAL for the reason mentioned, and detailed stripping was out of the question. As a consequence, maintenance likely took the form of simply swabbing out the bore and applying more oil to the carrier. This additional oil only served to attract more sand; the more oil, the thicker the resultant sludge mix once sand entered the picture. This sludge served to slow the speed of the carrier during cycling. This was likely the reason for the Israelis opting to try the grooved carrier design later on down the road.

In a desert environment, the oil must be kept to a minimum to help reduce the likelihood of attracting sand. This relatively precise method of lubrication was no doubt difficult to perform under the conditions faced by the Israelis during this fighting.

Some troops no doubt attempted to use dry lube like graphite powder, but the intensity of the fighting once again would have made this form of lube less effective. Graphite powder works in a pinch provided the weapon isn't being fired heavily. However, heavy use was likely the case for most FALs in use by Israel during the late 1960s. Graphite powder as a lube is better suited for limited use during arctic conditions, where the cold helps prevent the weapon from heating up excessively. The heat in desert warfare conditions makes the graphite less effective. The resultant heat buildup due to heavy use and the climate creates a condition of heavy metal expansion, which creates more friction during operation. Not only does this cause excessive wear, but the reduced clearance between cycling components only serves to "push" the graphite powder away from the parts it is meant to lubricate. In effect the weapon soon winds up running hot and dry.

Graphite as a lube in desert conditions is at its best when the weapon is being carried often but used only intermittently, as dry lube doesn't attract sand as conventional gun oil or grease does. Graphite under these conditions helps to keep the weapon clear of sand for longer periods yet does provide for adequate lubrication, provided the weapon doesn't heat up too much (minimal firing).

This was not the case for Israeli troops during this war and oil was required as the only suitable weapon lube under these conditions. This is why any excess of lube would have quickly led to malfunctioning in a short-stroke design like the FAL. That the AK fared better was partially due to its different gas system but also due to the greater clearance present in the AK design.

The UZI performed reliably due to both its large recess grooves and its blowback system of operation. As the UZI has no gas system, fewer parts of the weapon were at risk of accumulating sand and grit. Cleaning time for the UZI is also reduced due to its simpler method of operation.

In point of fact, the jamming issues of the FAL were not the sole reason for the Israeli's ultimate decision to replace the weapon. Most Israeli FALs in use were of the standard variety, with its 21" barrel and fixed stock. This made for an unwieldy rifle that was less than ideal for both urban combat conditions as well as transport vehicles. The much more manageable FAL

50–63 folding stock carbine model would have been a better choice for issue under these conditions, but this variation was not available at the time Israel adopted the FAL. As a result, it was not what Israel had produced and issued to her troops.

While the FAL had shown that any good design can have problems under adverse desert conditions, cold climates have proven equally detrimental to a weapon's performance. This is best illustrated by the U.S. Marine Corps' experiences during the Korean War or, more specifically, the battle at the Chosin Reservoir. The extreme cold during this fight caused failure of most U.S. small arms, with the M1 Garand faring somewhat better than most. The M1/M2 carbine series seemed to have regular trouble under these conditions of extreme cold.

The U.S. studied the M1/M2 carbine issue later and the primary causes for the failure were discussed earlier. While weak action springs and reduced gas pressure were officially the determining factors, one would be foolish to not consider the cold weather's effect on lubricants like oil and rifle grease. During the battle at the "Frozen Chosin," marines were forced to run their weapons dry whenever possible. It was rumored that thin hair tonic oil hardened less in the cold, and this oil was also known to see use as a substitute weapon lube. The traditional U.S. issue Lubriplate weapon lube would harden to the point of causing the actions of most U.S. small arms to freeze up and work so sluggishly that troops often found themselves firing manual repeaters. This was, of course, not a good situation to be in, so the jarheads left their weapons dry whenever they could.

Again, it is likely that the long-stroke gas action of the M1 Garand was largely responsible for its better overall reliability. In comparison, the M1/M2 carbine's short-stroke gas tappet system was more dependent on the initial momentum of its "operating slide." The well-used six-year-old recoil springs didn't help. Any hardened oil or grease present would have only served to seal the carbine's fate and all but guarantee that the weapon would fail to reliably function.

In contrast, the AK-47 performed better in general under such extreme weather conditions. This was most likely due to the added momentum of its heavy one piece carrier and gas piston, which was the result of its long-stroke gas system, in addition to the generous clearance present within its action. There were no doubt many other variables involved in the consistent superior reliability of the AK-47 under desert operating conditions when compared to the FAL.

The 7.62×39mm cartridge of the AK-47 likely contributed to the excellent overall performance of the rifle when compared to other contemporary designs. This cartridge is not known for target grade accuracy, but its geometry is ideally suited for self-loading weapons. It possesses a gentle shoulder angle and a tapered body that both contribute to better feeding characteristics than those of many other cartridges. Its short overall length also helps greatly in this regard.

By comparison, the much more powerful 7.62mm NATO is a much longer cartridge. By design, the two cartridges are almost complete opposites. The larger NATO cartridge is renowned for its excellent inherent accuracy, and while it has proven a reliable round for self-loaders, it does have a much sharper shoulder angle. Traditionally this type of case design requires greater force during the extraction phase of an automatic weapon's operating sequence. It is for this reason that many manually operated actions utilize a primary extraction phase. This is where the fired case is pulled from the chamber only a short distance, often when the bolt is being unlocked. This usually takes place through some form of camming action of the bolt. This allows the added leverage to ease the amount of force needed by the user. On many self-loading weapons, this primary extraction phase is not always feasible and, as a result, the case is pulled straight from the chamber during extraction. This rather violent extraction usually requires a fairly stout extractor, which is needed to handle the force of extraction without failing at this crucial stage of cycling. The FAL's use of the longer 7.62mm NATO case was likely not the cause of the reliability issues it suffered during the Six-Day War, but it could have been enough of a factor if the rifle was already heavily fouled with carbon and sand sludge mixed with oil. It must be remembered that the FAL was originally designed to use a much shorter case, shorter even than the 7.62×39mm Soviet cartridge.

While the FAL filled the NATO armories for many years, followed closely by the HK G3, it has nonetheless largely been replaced by smaller, lighter 5.56mm designs. While these "new" 5.56mm rifles may have already been serving for over thirty years in some cases, there had not been any large-scale, long-term desert combat during that time, other than in nations using AKs or FALs/G3s. Because of this and people's habit of often having short memories, the shortcomings of the smaller 5.56mm round had largely been ignored. The issue of a long-running combat operation in Afghanistan has opened the eyes of many to the shortcomings of the smaller NATO cartridge. With engagement ranges averaging greater than what is usually encountered, the advantages of the 7.62 NATO are being realized in an unpleasant manner. The British were longtime users of the FAL (L1A1) before switching over to the L85 Enfield series. This decision has had its drawbacks, to say the least. The limitations of the newer British bullpup led to a recent purchase of several hundred American designed 7.62mm NATO sharpshooter rifles (L129A1) for issue to designated British troops.[14] Should a severe need arise, the British will no doubt re-issue their old L1A1 rifles to deal with these long engagement ranges (provided they have maintained an inventory and not scrapped them all), as the U.S. has recently done with our own M14 (most of which are gone). We can only hope that current conditions in the Middle East cool and that combat in Afghanistan will be over before that need arises. If the L1A1 is needed again in a sandy environment, the Brits will just have to remember to go light on the oil and keep the sand out (easier said than done).

While the U.K. has replaced its version of the FAL, it continues to serve as standard or limited issue in other countries. It remains in production in Brazil at the IMBEL (Indústria de Material Bélico do Brasil) factory, and a high-quality civilian version is produced in the U.S. by DS Arms, Inc. In fact, some consider the DS Arms rifles to be the best FALs made to date. These civilian rifles are offered in the usual versions, in addition to some completely new and original variations. There is even an ultra-short pistol variation made. Though some adventurous people might be interested in firing this variation, one can only imagine the muzzle blast generated by such a short FAL.

DS Arms, Inc., began current business operations by selling FALs made from imported components, but after getting hold of Austrian prints and tools they began making FALs from scratch.[15] They currently offer a large variety of FAL models, from heavier match and sharpshooter versions to standard models to a variety of short barrel carbines and ultra-short rifles. Some of these require registration with the BATFE (Bureau of Alcohol, Tobacco, Firearms and Explosives), as they are considered NFA (National Firearms Act) weapons. Fixed or folding stocks are offered, as are receiver covers with built in rails. The fit and finish of these FAL rifles is often considered superior to military FALs. DS Arms uses the designation "SA58" for their FAL series, the "SA" meaning "semi-automatic" and the 58 from the Austrian version of the FAL, the StG 58.[16]

DS Arms offers more than a dozen variations of FAL. The basic model is the standard 21" barrel with either a fixed or folding stock, lightweight alloy lower receiver along with short pattern flash hider. Accessories available include carrying handle, upper receiver cover with scope rail, buttstock with offhand hold for bench shooting, and an M249 pattern pistol grip.

The 18" carbine barrel is the next in the DS Arms lineup and is available in the same configurations. These carbines are also available in 16" models. The Congo model comes standard with the carrying handle, though most variations have the carrying handle cutout in the receiver. The upper receivers are available in type 1 and type 2. The type 2 upper receiver has lightening cuts at the front and back of the receiver as with traditional military FALs.[17]

The folding stock is available in both the standard pattern that FN has used for years as well as a SOCOM folding stock that very much resembles the M4 type collapsible stock in profile. A great many other accessories are offered, including short and long muzzle brakes to be used in place of the flash hider for those who prefer less barrel displacement when shooting. Also available is a gas block with a built in rail section for the attachment of an accessory at this location.

Special ultra-short barrel versions are offered for law-enforcement and military agencies. Operational Specialist Weapon (OSW) models are available in 11" and 13" versions. These models come with a redesigned short gas system that functions with these extremely short barrels. The short gas system was designed to allow better functioning for barrels under 16". For barrels longer than this, the standard gas system will perform fine. These OSWs are offered in select-fire lower receiver variations for these units. This makes for a very potent 7.62mm NATO weapon that no doubt comes with a severe muzzle blast. Standard features for the OSW include a synthetic folding stock and a multi-rail forend to allow for the attachment of different accessories. DS Arms' use of alloy lower receivers is a valuable weight saving measure and helps keep the weight of these FALs to a reasonable limit.

FALs have also been available from some other importers. These usually use the Brazilian IMBEL parts, with the receiver coming from assorted sources in the U.S. to circumvent legal obstacles. Some of these rifles are of better quality than others. Parts kits from several sources are available for anyone looking to acquire a separate FAL receiver and build their own semi-custom rifle.

While the FAL was quickly proving itself around the world as a combat worthy rifle, no civilian models were originally available in the U.S. In later years this trend would change. Often, makers begin drooling over the mere thought of the U.S. civilian market before any military service has even adopted a new design. The first civilian FALs to enter the U.S. began to appear in the early 1960s. These early Browning imported FALs were modified to fire in semi-auto mode only. The Alcohol and Tobacco Tax Division (ATTD), which was the early version of the current Bureau of Alcohol, Tobacco, Firearms and Explosives, had approved this early FAL import for civilian sale. This decision may have been a bit hastily made, as it was later reversed due to the ease with which this rifle could be restored to its selective-fire condition.

These original import FALs had the auto sear removed and a replacement selector lever had been installed. This lever was limited to movement between the safe and semi-auto positions only, though the lower receiver remained unaltered. Finally, the trigger rebound plunger was changed to prevent the fully rear stage that is necessary for fire in the full-auto mode.

While these modifications prevented the rifle from firing in the full-auto mode, these parts could easily be replaced with the original components. Clearly this was a problem for a civilian rifle intended for sale anywhere in the U.S. Further modifications would be necessary to prevent the FAL from being returned to its selective-fire state.

These early Browning imports have historically been known as the G series, referring to the prefix on the serial numbers of these particular weapons. Later imported semi-auto FALs also had a "G" prefix, but the range was much higher on the serial numbers.

These later "safe" rifles had machining differences within the receiver to prevent any later installation of an auto sear. The later FALs that were imported into the U.S. lacked the full-auto position on the lower receiver.

Other changes were needed to make the FAL acceptable for import. In addition to the cutouts for the auto sear being left off, the two-position lever lacked the full-auto position. The ejector block was also modified for semi-auto only FALs.

These types of modifications are pretty standard for most semi-auto only models. This is because the two primary areas of concern on any selective-fire weapon, when converting to a semi-auto only model, are the auto sear removal and the carrier components. These changes are needed because these are generally the components that differentiate a full-auto from a semi-auto. The same areas of focus are present on the semi-auto AR-15 when compared to its military counterpart, the M16. The lower receiver and carrier are the primary parts that differ between the two models. The reason for this is that the auto sear must be activated by the carrier to make any rifle fire safely in full-auto mode. The carrier must activate the sear only after the bolt is fully locked (rollers engaged in the case of the G3). If any other method is attempted, the rifle may fire before the bolt has fully locked. This will lead to disaster for anyone near the rifle at the moment of firing.

With redesigned replacement components in these specific areas of the action, fully automatic fire is no longer possible. By merely removing the auto sear and altering the carrier design, the weapon is rendered incapable of automatic fire. In most cases, the geometry of the carrier and receiver is altered to discourage anyone from attempting to reverse the procedure.

While the FAL may have seen its heyday, it still serves on in reasonable numbers in some parts of the world, especially in Africa, where it was the preferred weapon (followed closely by the HK G3) for many years, until the arrival of vast quantities of AK-47s. It is also in limited service in several South American nations, with Venezuela only recently replacing it with the Russian 7.62×39mm AK-103, the modern version of the AK-47/AKM series. Argentina was a longtime user of the FAL (adopted 1955).[18] Production of the rifle took place near Rosario at the national armory, beginning in the early '60s and continuing for many years. Attempts to replace it with more modern 5.56mm native designs were not successful, mainly due to the economic conditions there. The FAL was used by opposing armies during Argentina's war with the U.K. in the Falkland Islands, with reports of British troops often preferring the captured full-auto Argentine model over their own semi-auto L1A1. It is likely that the FAL will continue to serve as Argentina's primary rifle for many more years due to lack of funds to develop a new rifle. Recent attempts to modernize the rifle would seem to be the most economical approach.

While the FAL has had its share of trouble in desert climates, it has served admirably in every other environment on earth, and the sheer numbers of this rifle made over the years combined with the long range ability of its 7.62mm NATO round make it the most viable option for any future desert operations involving those nations that still keep the rifle in their inventory.

The final portion of each of the following chapters will be a step-by-step rundown of the field stripping and maintenance procedures and tips specific to each rifle.

The FAL's gas regulator and its proper adjustment have been covered, so we will move straight into its field stripping and will also cover maintenance tips specific to the FAL. The FAL, like most military rifles, was designed to be what is popularly called "soldier proof." This meant that it was to be easy to disassemble and reassemble for cleaning and repair purposes. A "soldier proof" weapon also meant that in taking the weapon apart for cleaning there was little likelihood of losing, bending, or breaking parts during the process. It further meant that a weapon was designed in such a manner that there was also little chance of putting the weapon back together improperly and that the parts would only fit together in the correct manner during the reassembly procedure. The FAL was a well-designed rifle in all of these areas.

The first step in field stripping the FAL, and all firearms for that matter, is to clear the weapon to ensure that it is not loaded. With all box magazine–fed weapons, like the FAL, this is best done by first removing the magazine. On the FAL, the magazine release lever is a paddle type release, as mentioned. This lever is usually activated by pushing forward with the thumb while grasping the magazine. This method is standard for most box-fed magazine rifles. The paddle type release is by far the most common type of magazine release found on military rifles, so this technique will apply to most rifles as well. After the release lever is pushed forward, the magazine is then rocked forward from the bottom. This tilting motion allows for the magazine's frontal lug to be disengaged from the corresponding catch in the magazine well. The exact geometry of the magazine lug can vary somewhat depending on the particular variation of FAL (metric pattern/inch pattern), and as a result they are often not interchangeable. This means that British L1A1 magazines should not be used with a Brazilian FAL and vice versa. Once the magazine has been removed, the operating handle must be drawn to the rear to retract the bolt so that the chamber and magazine well can be visually inspected in order to verify that the weapon is indeed empty. Of course, this procedure can also be done by simply pulling back the operating handle and, if the magazine is empty, the bolt hold-open catch will automatically activate. This only applies to those FALs that have an operable hold-open device, however, as some variations do not possess this feature.

Once the FAL has been cleared, the next step is to allow the carrier to ride forward slowly until it has reached its return position. If the hold-open catch has been engaged, the operating handle will need to be pulled back before it can ride forward. At this point, the weapon should be placed on safe (if not already safed), leaving the hammer cocked. With the FAL, the safety can actually be applied before the hammer has been cocked and it is best to place it on safe before doing anything else. With many rifle designs, like the M16, the hammer must be cocked in order for the safety to be applied. The hammer is left cocked in order to prepare the carrier for removal. To accomplish this, the takedown lever is pivoted in a clockwise direction to allow the upper receiver to pivot upward as with a top-break shotgun. This break open will then allow the operator to withdraw the bolt carrier from the receiver by grasping the recoil spring guide and pulling the carrier and bolt assembly until it is free from the upper receiver.

Once the carrier group is out, the bolt can then be removed from the carrier by pressing in the rear of the firing pin until it clears the bolt carrier. Once the bolt is free from the carrier, the firing pin and spring can be removed by first drifting out the firing pin retaining pin using a small punch or even the tip of the bullet. Make sure the orientation of the firing pin is noted, as it is a characterized part and correct positioning is a must for proper reassembly. Once the firing pin and spring are out, the final step is to remove the sheet metal receiver top cover by sliding it off to the rear. The only remaining step for field stripping the FAL at this point is to disassemble the gas system.

The gas system is stripped by first turning the gas plug ninety degrees and pulling it forward toward the muzzle. Depending on the particular model of FAL, either a retaining plunger or lever holds the piece locked in place, and activating this lock is first necessary to allow the gas plug to be rotated. Once removed, the gas piston and piston return spring can be pulled forward toward the muzzle until they are clear of the gas block.

The forend panels can be removed by removing the retaining screws, but this should only be done if needed. Cleaning would be necessary if rust or mud was present underneath or if the rifle had accidentally taken a bath.

This completes the FAL field stripping procedure and no further operator disassembly is recommended by the manufacturer. Only a unit armorer should tear down the FAL any further.

Once the FAL is field stripped, the small total number of components gives clear evidence of the thought that went into the design of this rifle. The amount of forethought that went into the design of the FAL helps make the rifle easy to maintain in the field. Saive's work in this regard was apparently quite successful.

With regard to providing proper cleaning and lubrication on the FAL, less is often more in many aspects. Maintenance requirements often vary from one design to the next, with each design having its own particular needs, and proper applications often depend on operating conditions and climates. Weapons used in a desert theater of operations require different maintenance techniques than weapons being used in jungle or arctic environments. Luckily, the FAL rifle is a weapon that performs best with minimal lubrication. This is often the case with the better designs and lends itself to more reliable operation as well.

For cleaning, several solutions may be required, depending largely on what type of ammunition is being used. The corrosive priming compounds used in ammunition production that have been largely replaced in recent years are still encountered on occasion. This type of ammunition requires the user to provide special cleaning practices in order to prevent damage to a weapon. Hot fresh water is actually one of the best solutions for removing the corrosive salt deposits left behind by these older priming compounds. However, this water must be thoroughly dried before storing the weapon for any period of time. Regarding powder fouling and carbon deposits from the smokeless powder commonly used today, any of a number of gun cleaning solvents will be effective. Common solvents from kerosene and gasoline to acetone have been used over the years with no harmful effects to firearms provided they are wiped clean before use or storage. Solvents containing ammonia are effective at removing the copper deposits that develop from the jacket material used in many modern bullet designs. There is often concern regarding ammonia and its use on plated metals, such as the chrome plating often found on many barrel interiors on military weapons. This issue is largely dependent on the plating process. Some solvent manufacturers do recommend that nickel plated weapons should be wiped clean of any solvent or damage can occur. This is due to the fact that some plating techniques utilize a copper preparation on the metal before the plating is applied. The concern here is that the ammonia in the solvent can dissolve the copper and caused the plating to peel away. This does not appear to be a major problem these days, as many modern plating techniques use different methods that are devoid of any copper preparation on the surface metal. As long as the solvent is completely dried before the weapon is stored, no problems are likely to occur.

After cleaning, a light coat of synthetic gun oil is a good idea to protect the bore. Many solutions exist today that provide not only a cleaning action but lubrication and preservation as well. These triple action solutions are becoming quite popular for their convenience. This is one of the more common military cleaning solutions used today.

The bolt face of any self-loading weapon should be thoroughly cleaned and the area under the extractor is of particular importance. If a plunger type ejector is present as with the U.S. M14, it should also be cleaned and lubed.

Nylon brushes are popular for cleaning, though brass or bronze brushes provide more aggressive cleaning action. Stainless steel brushes are generally used for highly aggressive cleaning on extremely dirty areas of the weapon. Stainless steel should only be used when needed and only if deemed safe by the manufacturer. Regarding aluminum components, only nylon should be used here. This is a universal rule, as any of the other brush materials will quickly scratch the aluminum alloys no matter how well they have been anodized or hard anodized.

The gas piston area of the FAL should be kept dry, with no lubrication of any kind needed. This is often the case on many gas operated weapons, but not all. Cleaning the piston area should be done with a nylon brush or solvent soaked patches and plenty of elbow grease, as scouring materials will cause undue wear on these close tolerance components. Excessive wear in this area will lead to reliability issues over time. Pipe cleaners have long been popular for cleaning gas ports in this area as well. Brass or bronze brushes are OK for really stubborn fouling if a nylon brush is not up to the task.

For proper lubrication of the FAL, only the metal surfaces that show friction wear should receive oil or even high-temp bearing grease (long a popular technique for competition shooting with the M14). Examples of such areas on the FAL would be the bolt locking lug area and its corresponding recess area in the receiver. The receiver guide rails are other areas that tend to show friction wear. For springs and pivot pins, a light application of gun oil with minimal excess is often the best approach.

That is all that is necessary to keep the FAL up and running. With good cleaning and minimal lubrication levels in the key areas, the FAL will function reliably under most conditions, with extra care and attention to keeping the weapon clear of any sand in desert conditions being of utmost importance.

Reassembly of the FAL is the exact reverse of the takedown procedure. This is usually the case for most takedown applications regarding any weapon. The only variation in design that changes the field stripping procedure is on the folding stock models of FAL, where the recoil spring is housed within the carrier itself and the long "rat tail" return spring guide used on the fixed stock variation is not present. This does not, however, alter the order of procedure. It merely adds the step of removing/replacing the recoil spring within the carrier spring recess.

Regarding other maintenance techniques, it is a good idea to keep the sights clean, especially the rear sight. Having this obscured at the wrong time could prove a fatal bit of neglect and oversight.

These maintenance techniques will apply equally well to civilian FAL variations as well as military models and, if followed closely, will guarantee a reliable operating rifle, which, over the years, has come to represent the very word "NATO."

3

HK G3

While the FAL was undoubtedly the most popular post-war Western rifle, it was followed quite closely by the German HK G3. These two rifles represented very different technologies yet achieved the same end result. The FAL was a gas piston–operated rifle and utilized an expensive forged (later cast on some models) and machined receiver. The G3, by comparison, primarily used stamped steel components, including the receiver, and had no gas system whatsoever. Instead the HK G3 relied on a roller locking system first developed from the wartime German Mauser designed StG 45. It is worth noting other similarities between the G3 and the StG 45. The method of stock attachment and recoil spring location are quite similar, as is the modular trigger group concept. Since the designers of the G3 were largely former Mauser people, similarity between the two rifles is quite natural. The advantage of the technology used in making the G3 is that it is cost effective and requires relatively low-tech manufacturing ability. This came about due to the StG 45's development under wartime conditions. The Germans needed a weapon that was cheap and easy to produce and required little in the way of highly skilled machining, though a few components still required conventional machining. The G3's ease of production ensured that setting up manufacturing facilities in developing nations was a fairly easy task. It was primarily for this reason that the G3 was nearly as popular as the FAL for many years. It was an attractive alternative to the FAL for nations that wanted a rifle that could be produced locally and yet did not want to spend too much to do so. Many of these nations had a small pool of skilled workers from which to draw, yet unskilled workers were abundant and could be trained in short order to operate many of the machines used in the G3's manufacture. While this may lead some to the conclusion that the G3 was a cheap second choice to the FAL, it was anything but. The G3 has long

A sharpshooter's version of the HK G3, the G3A3ZF. This model uses a modern tactical scope atop the usual HK claw type mount. A cheek piece is also fitted to allow for easier scope/eye alignment and a mounting rail has been fitted to the forend along with a Harris pattern folding bipod. The effective range of this rig far exceeds that of any 5.56mm system. (MC2 J.D. Chandler)

been known for its excellent accuracy and reliability under extreme combat conditions.

As I briefly covered the history and evolution of the G3 in the first chapter, I will go right into its design aspects. While the G3 is less expensive to manufacture than the FAL, it is a very well constructed rifle with a good reputation for longevity. This is largely due to its quality of design. As mentioned, the receiver is a stamped steel component, but this part primarily serves as a guide for the bolt assembly during recoil.

The area that absorbs the brunt of the firing stress is the barrel trunnion. This is just another word for a barrel extension as used in many other modern military designs, which utilize either stamped steel or forged aluminum receivers, such as the AKM variant of the AK-47. In the case of the G3, the trunnion is spot welded into the receiver and houses the recess that holds the bolt locking rollers during firing. As those familiar with the HK delayed-blowback system know, the barrel chamber is cut with longitudinal flutes to aid in the extraction process. The barrel itself is press fit into the trunnion and then pinned in place.

The bolt of the HK is comprised of several components. These are the bolt head, the locking rollers, the locking piece and the bolt body (carrier). These components make up the heart of this system. As the pressure builds in the chamber upon firing, the cartridge case is forced rearward. This pushes the bolt head back as well. As the locking rollers are set in their recesses, there is a locked bolt state of sorts. The rollers are being forced back along with the bolt head at the same time as the rollers are striving to climb up the curved surface of the recess. As this happens, the rollers are also acting on the angled locking piece. In a short time, the rollers force the locking piece backwards at an accelerated speed. This forces the bolt body rearward at a fast pace and it takes the rest of the bolt along with it after a short distance of free travel. Once the locking piece has moved to the rear, the rollers are allowed to retract and the bolt is free to move to the rear. This unlocks the bolt, in a manner of speaking. As the whole process is unfolding, the bullet has left the barrel and chamber pressure has dropped to a safe level by the time the bolt head moves to extract the empty case. This delayed bolt head movement accounts for the "delayed" in the term "delayed blowback." In fact, the breech is never fully locked. This action type results in rather harsh extraction, hence the need for a fluted chamber to help prevent cases being ripped apart during the operating sequence.

The G3 trigger system is a modular unit, as mentioned. Some users have complained that the G3's safety lever is awkward to manipulate and poorly located, requiring users to move their hand from the pistol grip in order to operate the selector. As a result, HK eventually offered a modified safety lever that was easier to manipulate. Later versions of HK trigger groups offer not only semi- or full-auto options but also either two- or three-round burst options as well. Some of these trigger groups offer semi- and burst only. These trigger groups are usually seen on the MP5 submachine gun, which uses a similar roller locking system. The G3 uses a conventional coil type return spring and guide, which are attached to the buttstock. This buttstock can be either fixed (G3A3) or collapsible (G3A4). There was even a side folding stock that was developed in later years, but this is seldom seen on service models. As with other HK delayed blowback long arms, there are mounting points for optical sights and HK has offered an excellent quick detachable claw mount for many years. This is a high-profile mount to allow for optical clearance of the front sight post.

The iron sights on the G3 have evolved over the years. The original version used a simple flip type aperture, while the modern rotary aperture sight has been the standard for quite some time now. This sight has proven rugged and easy to adjust for different range settings out to the maximum effective range of iron sights, which is usually regarded to be between four and five hundred yards. The maximum range setting on the G3's turret is 400m. The rear sight can be zeroed using a special sight adjustment tool or a combination of conventional tools can be used, for someone without access to the purpose built tool. The front sight is a post protected by a circular hood. This post, as well as the rear sight turret, can be replaced with variants containing radioactive elements for night shooting if desired.

The G3 has two magazine release options. One is a push button located on the right side similar to that of the M16, while the other is a paddle lever located behind the magazine. This paddle was not present on the civilian HK model 91 variants, as a shelf was welded in this area to prevent installation of full-auto trigger groups.

On the G3, the trigger group allows for semi- or full-auto and has a release lever that must be activated before the hammer can fall. This is similar to many selective-fire weapons and is to ensure that the bolt body is fully forward and the locking rollers have been engaged in their recesses. In addition, a locking lever is present to prevent the bolt head from rebounding off of the breech face after a round is chambered. This lever is released once the bolt body is fully forward. This system is different from that of the HK MP5 submachine gun, which uses tungsten added to the bolt for increased mass, which accomplishes the same result.

As the G3's trigger is not known for its light pull, one would think the rifle was not usually used as a sniper weapon. Quite the contrary; there have been several sniper variants made over the years. The G3A3Z was one of the first sniper versions of the G3. This was primarily a standard G3 with a scope (usually a Zeiss) and mount fitted to the receiver. A more specialized sniper variant was the G3SG/1.[1] This model used a select grade barrel and a special set trigger group. When the set lever was pressed, the trigger pull weight and travel were reduced to an absolute minimum. This model used the wide plastic "tropical" forend and the HK folding bipod. In addition, a cheek piece was fitted to the stock to help with scope/eye alignment by raising the location for the user's cheek. The selective-fire option remained on both sniper variants, unlike many sniper versions of battle rifles, which often offered the semi-auto option only. This was often done to prevent accidental full-automatic fire.

One advantage the G3 has regarding its trigger mechanism is that it is a modular pack that can be quickly and easily replaced if any parts should fail or break. While the G3's trigger is known to be heavy and not overly crisp in let off, it is a rugged design that has proven more than field worthy. The trigger housing design was changed in the '90s and is now a single plastic unit. The original was a metal housing with a plastic grip fitted to the frame.

The trigger pack itself fits into the housing and is secured in place by the arm of the safety/selector lever. To remove the trigger pack from the housing, all that is needed is to pivot the safety lever to 12 o'clock and slide it out. The trigger pack can then be withdrawn from the top of the housing.

While most battle rifles closely resemble the look of their civilian counterparts, there are several important internal differences. These are due to changes in design that are necessary in order to prevent the easy conversion to a selective-fire weapon. Regarding the differences between the original G3 and the HK 91 (though the PTR 91 is similar), there were several alterations made to stall any conversion attempts. First up is the small "shelf" welded over the paddle release location (the reason for only the push button mag release option). This shelf not only does away with the paddle release lever but also was placed to block the location of the forward trigger group mounting pin hold of the selective-fire G3. This was to prevent someone from simply switching trigger groups and creating an illegal selective-fire rifle.

In addition to this major obstacle, the semi-auto trigger pack itself is different, as is the entire trigger housing, which has no front pin hole and uses a "lip" to secure its forward portion, rather than a mounting pin.

The final component that is altered is the bolt carrier. On the selective-fire G3, the bolt carrier has a ramp at its rear section, which is needed to trip the full-auto sear release. The semi-auto only bolt carrier has a slot cut where the ramp would otherwise be so that this action cannot take place even were the full-auto trigger group present.

The end result of these modifications, aside from the loss of the paddle magazine release option, was that no conversion could be easily done to the rifle. Of course a machinist with the proper parts could perform this with relative ease. This would, however, carry with it a rather stiff penalty and few machinists will do this type of work. This is not to say that many machinists have not been approached with such requests on more than one occasion.

Even the standard G3 rifle was capable of excellent accuracy. The G3 has long been known for its quality cold hammer forged barrels and polygonal rifling. This form of rifling is said to reduce bullet deformation along with offering increased muzzle velocity and reduced barrel wear.

The G3 uses a twenty-round standard magazine like most 7.62mm NATO rifles. These magazines were available in either steel or aluminum. The aluminum magazines are reportedly reliable but offer a shorter service life. After significant use, the feed lips on the aluminum magazines are more likely to fatigue and bend or crack. A fifty-round drum magazine is available, but this option would seem to offer poor handling characteristics and would make the rifle extremely unwieldy and heavy. Traditionally, drum magazines are also more fragile and prone to malfunction than box variants.[2]

An excellent folding bipod is available for the G3 and was standard on the SG/1 sniper variant as mentioned. The forend used on this variant, and on many later models, had recesses molded into its sides to help house the bipod when in its folded position.

Takedown of the G3 rifle is easy and the weapon is quickly broken down into its major subassemblies. The cocking handle is located on the left side of the cocking lever housing, which is the long tube that lies above the barrel. This is often mistaken for a gas cylinder tube. There is a notch in this housing to allow the bolt to be locked in the open position. There is no last round hold-open device as with the U.S. M14 and most FALs. The cocking lever folds along the lever housing to help prevent snagging. The cocking handle remains stationary during firing. One particular advantage the G3 has over its contemporaries is that, as it lacks a gas system, there's less chance of jamming when heavily fouled. This also means a shorter cleaning time for the G3 than that of gas-operated rifles. One area that must be kept clean is the locking roller recess area, as any excess of grime built up in this area may prevent the locking rollers from fully engaging, which will prevent firing. This is actually one of the G3's safety systems. If the rollers have not fully opened into their recesses, the firing pin cannot move forward far enough to fire a chambered cartridge. As mentioned earlier, the G3 has proven an extremely rugged and effective service rifle since first introduced in the late 1950s. It has not been adopted by quite as many nations as the FAL, but its lower production cost and less time-consuming manufacture have made it an extremely popular rifle worldwide. It has been produced by a great many countries, with production continuing even today in some locations. Pakistan and Turkey still offer the G3 as a production rifle and it is made in several different versions, like the collapsible stock model, in addition to the standard fixed stock variation.

The HK G3 is one of the few rifles to be adopted as a standard service weapon that did not utilize some form of gas operation. The SIG StG 57 and the rest of the 510 series did see some service, but this was largely a design that stayed home in Switzerland. The G3, however, was one of the most common service rifles to be seen in the armories of the world in recent years. It could never have achieved this status if it were not a completely battle-worthy design.

Overall the G3 had few weaknesses. It did offer rather harsh recoil due to the nature of its design, but this is only of minor concern, as all 7.62mm NATO rifles are more severe in their recoil than smaller 5.56mm and 7.62×39mm weapons. Also, the G3 lacked a last round hold-open device in addition to being slightly heavier than the FAL and M14, though the difference in weight was only about half a pound. The collapsing stock G3A4 version was somewhat heavier.

In the G3's favor, it proved to be one of the most durable and dependable service rifles, even by today's increased standards of production. It also was usually agreed to be more accurate than the FAL and equal to the M14 in this regard.

If the G3 did have an Achilles' heel, it was its sheet metal cocking handle tube and receiver. These were stamped components, which were designed to be cost effective yet serviceable at the same time. The issue is that these parts were designed to be relatively thin in cross section in order to save weight while providing adequate strength for service. The result of using this metal for such large and exposed components is that they are prone to deformation from hard impact. If the G3 is bumped against a hard object with enough force, dents will be the end result. If the dents are severe or deep enough, reliability will be affected. In field service, a rifle bumping hard objects is not only probable but inevitable.

A deep enough dent in either the G3's receiver or cocking handle tube can slow the carrier's speed during operation due to the increase in friction. HK quickly became aware of this issue and as a result designed tools to correct the situation. Straightening mandrels for the G3 were issued to service armorers to allow them to repair the damage to any fielded weapon that suffered this affliction. The armorer's manual for the G3 even illustrates the procedure as part of routine repair work when the G3 is brought in for overhaul.

While any gas-operated weapon can also be adversely affected by severe impact, the impact must normally be far more severe to affect the reliability of the weapon. This is due to the designs of most gas systems, which are generally far more heavily built and less susceptible to impact damage. The longer, more fragile gas tubes and operating rods of most gas actions are protected by the forend in most designs. As a result, these are generally not damaged during normal service.

Oddly enough, one gas-operated design suffers a weakness similar to that of the G3, and this model is none other than the AK-47/AKM/AK-74 series.

The gas tube of the AK pattern rifles (including the Israeli Galil) is an exposed sheet metal component that sits atop the barrel and is quite susceptible to damage from any direct impact. Fortunately for the user, this is a part that can be easily switched out for a replacement, the whole procedure taking no more time than that required to field strip the weapon (less than a minute for an experienced user).

Unfortunately, this is not the case for the G3. If a G3 suffers such damage to the receiver or cocking handle tube, the result can be an out of action rifle or, worse, a dead soldier due to a failed weapon. Damage severe enough to cause a jammed G3 rifle is fairly easy to spot by a quick rifle inspection and in fact is usually obvious at a glance. The needed repair is fairly quick to perform, but it does take a G3 out of service for a short time. The repaired rifle must also be tested for proper functioning before the armorer can return it to service.

The AK has a similar weakness regarding its sheet metal receiver cover, but like its gas tube, this can be quickly replaced, and function testing is not required, as the receiver cover serves primarily as a dustcover for the operating components. In fact, replacing the receiver cover on the AK can be done in a matter of seconds.

While this is a relatively minor issue regarding the service of the HK G3, it is a problem from time to time given the rough conditions within which a service rifle is required to operate. As the G3 is still on active duty in several nations, it is clearly not a major issue.

Aside from this matter, the G3, in general, requires less maintenance than most service rifles, and many familiar with the model consider it second only to the AK with regard to reliability. It was clearly a serviceable weapon, as it remained standard issue for West Germany from 1959 until finally it was replaced by the G36 5.56mm short-stroke gas-operated rifle in 1997. Since then, both German rifles have seen service in Afghanistan with favorable results. The decision to replace the G3 has, however, become an issue within some circles of the German military who feel the G3 was a superior rifle. This is certainly true in terms of power and effective range. In fact, German sharpshooters are still seen in Afghanistan carrying the G3A3Z sniper rifle.

There have been recent reports in the German media that claim the G36 is not all it was cracked up to be. To be fair to HK, it should be considered that this new rifle is not and never will be a long range performer. The same can be said when comparing the U.S. M16 to the older M14.

Civilian semi-automatic only models are available in the U.S. The PTR 91 is the most recently manufactured version. HK did offer the model 91 semi-auto during the 1970s and '80s. This was always an expensive model and sales likely suffered a bit as a result. This high price was surprising for a rifle that was designed to be inexpensive to produce.

The current PTR 91 semi-auto rifle is often considered to be of even better quality than the original HK model from Germany. This rifle's maker, PTR91, Inc., began operations (under a different name) using original Portuguese G3 tooling. Portugal was a longtime manufacturer of the G3, and there it was known as the m/961. Century International Arms, Inc., also made a semi-auto version for several years.[3] This model can vary in quality, but some have claimed that they shoot fine. For a time, this model was made

The primary reason for the battle rifle's return to service, the U.S. M4 carbine, seen here in its basic variation with no attachments ("vanilla"). The carrying handle on the M4 series is detachable and beneath it is a rail system that facilitates the mounting of any number of optical sighting options. While light and easy to handle, the M4 has an effective range limitation, which has proven to be a severe problem during the long range engagements common in Afghanistan. (MC1 Keith Jones)

using a stainless steel casting rather than the carbon steel stamping normally used for the receiver.

The PTR 91 rifles are offered in a great many variations, as are the DS Arms FALs. The basic model is the 18" fixed stock PTR 91F. This is the semi-auto equivalent to the classic HK G3A3. This same rifle is offered in a 16" carbine model (PTR 91KF) that still offers decent muzzle velocity. This same barrel length is available with the classic HK pattern collapsible stock. There is also an M4 style collapsible stock available. Rail forends are offered, as are original HK G3 style slim or wide "tropical" pattern forends. The PTR 91 receiver will accept the original HK pattern claw type scope mount, though some minor fitting may be required. The PTR 91 is not an exact copy of the HK G3, but is extremely close—close enough that most of the accessories will fit either design including surplus G3 magazines.

While the original PTR 91 rifles used large numbers of surplus G3 parts, supplies tend to dry up rather quickly and the current models utilize more new PTR 91 Industries–made components. This is not necessarily a bad thing, as the manufacturer can supervise the production of new parts to help ensure overall quality, as opposed to relying on parts that originated from other sources.

The current line of PTR 91 rifles very closely resembles the G3 and at even a short distance differences between the two are hard to spot.

At the heart of the G3's design (aside from the roller-locking action) was its low production cost. This was the stimulus for its design in the first place as a distant relative of the Nazi era StG 45. The original advantages stayed with the design to the present day. The roller-locking system was intended to do away with the added time and cost of having to machine a gas system in addition to the rest of the rifle. It also meant reduced overall weight in addition to lower cost and shorter production time.

This is likely the reason for the G3's continued production in nations such as Pakistan, Turkey and Iran, in addition to some African nations and Saudi Arabia.

This lower-cost production may help explain why the current PTR 91 semi-auto civilian rifles offer U.S. gun collectors an extremely accurate battle rifle look-alike for around $1,000, while similar counterparts to the FAL and M14 are going for considerably more.

PTR 91, Inc., is continuing to offer a greater variety of these rifles. Two of the options offered by PTR 91 were not available on the original G3. These are the rail equipped receiver and forend, though some current G3s in service have had these accessories retrofitted, and the M4 style telescoping stock. The M4 style stock is a nice feature for shooters who are more particular about the quality of check weld one can attain while shouldering a rifle. The original HK pattern collapsing stock was not well known for its comfort level when resting against the side of one's face, though some users didn't seem to mind much.

The real plus is the built-in rail system which makes for much easier mounting of optical sights. With the original HK claw mount, first the mount must be adjusted for proper tension and then the appropriate base can be installed. HK still offers its original claw mount, though an M1913 "Picatinny" rail is now the standard base (which is an improvement). While rugged and easily removed, the claw mount raises the point of scope/eye alignment so high that a cheek rest attachment is often needed atop the stock. This is not easily achieved if the original HK collapsing stock is in place. With a fixed stock, it is not much of a problem. Adding accessories, however, will rapidly increase the weight of an already heavy rifle.

Anyone looking to buy a semi-auto version of the G3 for sport shooting will likely want to save the brass for reloading, as semi-autos tend to have a huge appetite for rounds. Brass catchers are available for this rifle. These devices have long proven useful for rifles intended for use from helicopters as well (the brass rolling around on the floor of the chopper can be a hazard). The brass does come out of the G3's chamber with ugly dark lines along its surface, but they are still perfectly safe for reloading. In some cases the brass does get dented around the case mouth area. Depending on how bad this is, it may be best to discard the dented cases.

While the G3 has been replaced in German service by the new (mid '90s) HK G36, a 5.56mm short-stroke gas-operated rifle, it continues to serve on in other nations and remains in production among some of those. During the first Gulf War, the G3 reportedly served well with no reliability issues, despite being a thirty-year-old design at the time. It has recently seen considerable combat duty in Iraq and Afghanistan as well. It has been seen in the hands of coalition troops and some of these have been fitted with modern Picatinny rails mounted on both receiver and forend. While it is often considered to be the poor man's alternative to the FAL (a reputation gained from its use during the long Rhodesian Bush War), it is arguably the most durable service rifle made in this powerful caliber.[4] Due to its relatively low cost and ease of production for cash-strapped or developing nations, it will likely remain in service and production for years to come.

Field stripping the HK G3 requires, as all firearms do, first clearing the rifle to ensure that it is empty. This means placing the weapon on safe, removing the magazine, and visually clearing the chamber by pulling the operating handle to the rear. After the chamber and magazine well have been verified to be empty, slowly allow the operating handle to return to position, leaving the selector placed on safe.

The next step is to remove the two takedown pins from the left side of the receiver and place them in their respective storage holes located within the buttstock in order to prevent loss of the pins.

Next pull the stock assembly to the rear and clear of the receiver. The trigger group will then swing down by its front pin (front pin in trigger group is only for full-auto versions). The civilian trigger group can simply be pulled down and back to free it from the receiver. The operating handle stays within the tube located above the barrel. Holding the receiver with the muzzle angled slightly upward, pull the operating handle to the rear until the bolt assembly can be pulled out from the back of the receiver. It will usually fall free on its own. Next hold the bolt carrier assembly and turn the bolt head clockwise roughly ninety degrees until it can be pulled forward and clear of the carrier. To complete basic field stripping, remove the locking piece by also twisting and pulling free and then pull the firing pin along with its return spring, taking notice of the locking piece orientation for reassembly purposes.

Reassembly of the HK G3 involves simply reversing the procedures, with crucial steps being to ensure that the locking piece is correctly aligned and that the bolt head is not pushed too far into the carrier during the procedure. Correct reassembly of the locking piece merely involves making sure that the lug at the rear of the locking piece is lined up with the rounded portion of the bolt head. The second key step simply involves gentle handling of the bolt head during the reassembly process. In the event that the bolt head is pushed too far to the rear, the rollers will open and you will have to first retract the rollers before continuing any further. This usually involves nothing more than dropping the carrier backwards down the rear of the receiver. This action should force the rollers back into the retracted position. Use a plastic mallet or other non-marring device to help if necessary. Start from where you left off once the rollers are back into position.

Provided the roller recess area is kept clean and free of any excess buildup and that the chamber is cleaned as needed, the HK G3 can easily handle any and all battlefield conditions likely to be encountered anywhere on earth. It has proven this ability on more than one occasion during its heyday as a battle rifle.

Regarding maintenance issues with the G3, it was designed as a rugged, easy to maintain service rifle. While the FAL is a conventional short-stroke gas-operated rifle, it does have a gas system that must be cleaned in addition to the barrel and receiver area. However, the HK G3 operator has the added advantage of not having to deal with the extra time involved in cleaning not only the barrel/receiver area but the gas system as well. This is quite fortunate, as a gas-operated weapon has a tendency to get dirtiest in this particular area.

With the HK roller-locking action, there is no gas system with which to deal. The roller-locking action does have particular areas of concern, however, with regard to proper cleaning. The G3 and all other models in the original HK lineup were designed with chambers that had longitudinal grooves around their circumference. These grooves were necessary to prevent the cartridge case from sticking to the chamber wall during the extraction process. This was required due to the delayed-blowback action, which depended entirely on the force of the expanding gases to overcome the mechanical disadvantage of the rollers in their recesses.

While this operating system proved very reliable over the years, the chamber flutes allow a small amount of gas to float the cartridge case, which helps ease the force necessary for extraction. During firing, however, the chamber area tends to become extremely dirty, as carbon deposits also flow back during the firing process.

This area is important to clean properly, as a clean chamber is a necessity in any self-loading weapon but of particular importance in the roller-locking mechanism, as this is the key to the operating system itself. Fortunately, chamber brushes are made for just this purpose.

Aside from the chamber area, another crucial part of the action that demands special attention during cleaning is the recess area for the locking rollers. An excess of carbon buildup in this area can lead to the rollers failing to open completely. While this is not a safety factor, it can lead to a failure to fire, and at a most inopportune time. This was one of the safety systems designed in the roller-locking action. If the rollers are not fully opened, the firing pin cannot move forward to strike the primer.

While in the HK G3 a chrome lined barrel was not utilized, this has never posed any problems with stuck cases during extraction, most likely due to the addition of the chamber flutes.

Aside from keeping these specific areas of the action clean, no out of the ordinary cleaning or maintenance is required. Once the bore is clean and dry, a light coat of oil or preservative is recommended, as with all barrels. Again, as with all barrels, the oil should be swabbed dry before firing.

Regarding the other areas of the G3 action, just a thorough cleaning with solvent followed by a light coat of oil in the proper locations is all that is needed. This primarily applies to springs, pivot pins and any other areas subject to metal-to-metal friction.

If this basic maintenance regimen is applied to the roller-locking rifle system, it will function well for many years. In fact, the durability and reliability of this design is usually rivaled only by that of the AK-47.

4

U.S. M14

While the M14 had been in development roughly since the final days of World War II, its final version wasn't officially adopted until 1957 and first issue didn't take place until almost 1960.[1] The ironic part of the story is that the program was canceled in 1963. The last part of the story had more to do with political infighting than anything else. As the historical aspects of the M14's development are far too heavily laden with political infighting, we will move into discussing its technical characteristics and tactical ability as a battle rifle, as that is within the intended scope of this book.

While the M14 was adopted by the most powerful nation in the world at the time, it was the least popular of the three primary battle rifles of the post-war era. Part of this may have been due to its cost, especially when compared to the G3. However, it is at times considered that the M14's old school looks and lack of a modern pistol grip placed it third. It is generally considered to be the least controllable of the three rifles when fired on full-auto, though many M14s were relegated to semi-auto fire only. Again, its conventional stock design played a part in this lack of controllability when fired on full-auto. As for production costs, it must be remembered that the FAL's receiver was also initially machined from a steel forging. In addition, the FAL's gas system was far more evolved and complicated. One of the reasons the M14 was chosen over the FAL by the U.S. military was its fewer total number of parts used in construction.[2] This was an original selling point. The FAL and the G3, however, had undeniably modern looks, especially when fitted with synthetic furniture, which was available almost from the start on both models. This no doubt helped drive sales for both rifles as well. While the FAL is often regarded as having the best balance and handling characteristics of the three, the M14 is no slouch in this regard, either. The G3 in comparison is often considered to be a clumsy rifle with regard to fast

The U.S. M14 in its original basic variation with exception of a wood stock (which has seen better days). While this model has only iron sights to guide it, it is still more than capable and can lay down effective fire to more than double the effective range of the M4 carbine. (SSGT Ryan Crane)

handling ability, especially with its awkward-to-manipulate safety lever.[3] While the FAL's heavy barrel SAW version did see some service, the M14's equivalent, the M15, was declared obsolete almost immediately and is never seen in the hands of troops; instead, the M14A1 took its place.[4] This was really nothing more than a standard selective-fire M14 with a modified pistol grip stock and some other minor changes. This stock was very unorthodox in looks and was designed to provide for more in line recoil force. A folding forward pistol grip and a muzzle break were also fitted to help in this regard. This model was never able to achieve any real success as a SAW for reasons similar to the FAL SAW models, specifically the lack of a quick change barrel. To be fair, the FAL SAW models did have some reliability issues in addition to the other shortcomings. The standard rifle version of the M14 may have been criticized as looking too old-fashioned and being almost impossible to control in the automatic mode, but it quickly earned a reputation for being one of the most, if not the most, accurate service rifles ever encountered. Certainly anyone with half a brain would not want to be anywhere near the receiving end of this rifle when it was in the hands of an expert marksman. The 7.62 NATO, the cartridge for which it was chambered, has also long been known for its excellent inherent accuracy. The same chambering no doubt played a small part in the better than average accuracy of the FAL and the G3 as well, with the FAL's accuracy limitations being more likely due to its tilting bolt locking system and the G3's shortcomings due mostly to its heavy trigger pull. The trigger system on the M14 is often praised for its quality of pull and crisp let-off. The trigger design was nearly identical to that of the earlier M1 Garand. The gas system of the M14 is also considered to be well designed and offers a gentler push than the harsh punch of the G3, though on the M14 the gas system is self-regulating and not adjustable, as is that of the FAL.

When the decision was made to adopt the M14, it was understood that much of the tooling to be used in its manufacture was the same as was used in the making of the M1 Garand, its predecessor. This was not the case, as it turned out, and the M14 suffered a great many setbacks during its early years of production.[5] One particularly well-known issue was the receiver failures of some early Harrington and Richardson (H&R) M14s.[6] While each manufacturer of the M14 had its own production problems, this was probably the worst. The final M14 contractor, Thompson Ramo Woolridge (TRW), Inc., is generally credited with turning out the best quality rifles. Springfield Armory (the former U.S. national armory) and Winchester also made their fair share. The receivers of all M14s were constructed to last for several hundred thousand rounds.[7] The receiver life was far in excess of the life of the barrel itself. Most barrels tend to be shot out after ten thousand rounds, with modern cold hammer forged barrels often lasting twice this. The jacketed bullets used on military ammunition tend to wear barrels out much faster than lead bullets, which are popular among sport shooters due to their lower cost. As mentioned in Chapter 1, Taiwan took possession of much of H&R's tooling and continued on with M14 production (as the type 57).[8] They reportedly made over a million of their own. Communist China made quite a few as well. Poly Technologies Corporation (Polytech) and China North Industries Corporation (Norinco) have both produced semi-auto copies primarily for commercial sale. Over the years, the primary source for semi-auto versions of the M14, however, has been Springfield Armory, Inc. (not the former U.S. national armory). This firm has been involved in production of semi-automatic M14 look-alikes since the 1970s. These differ primarily in the construction methods used for the receiver. The commercial Springfield is known as the M1A and uses an investment casting for the receiver rather than a forging. There's another quality semi-auto version being offered today from Fulton Armory. This model also uses a cast receiver, though the price is quite a bit more than that for the standard model from Springfield Armory, Inc. Both of these rifles are quality weapons for someone looking for a semi-auto copy of the M14. Fulton Armory uses actual U.S. G.I. parts for most of its components. As the supply of these parts is drying up, the cost is considerably more for Fulton rifles. Several variants are offered by both firms, though Springfield Armory weapons are far more common.

In addition to the standard model, Springfield Armory also offers an 18" barrel Squad Scout model with a short rail fitted on top near the rear of the hand guard. They also offer the SOCOM and SOCOM II 16" models with good handling characteristics and even more rails fitted (SOCOM II). Stocks for these weapons are usually synthetic and are available in different colors. Sniper and match variants are available from both manufacturers as well. Match grade barrels are available from different makers and are available in stainless steel, depending upon the model. These match grade barrels are heavier than the standard barrel and this will add to the overall weight of the rifle. Also, it must be remembered that changing components can affect the fit of other parts if one is attempting to build a custom rifle. Oftentimes changing one part will require the replacement of several components due to dimensional differences.

As for the M14's inner workings, it uses a short-stroke, self-regulating gas system. What this means is that the gas system cuts off the supply of gas to the cylinder once enough volume had been delivered to operate the action. This design was a departure from the M1 Garand, which uses a long-stroke gas system. The outward similarities between the M14 and the M1 Garand are obvious and the M14 is often considered to be nothing more than a product improved M1 Garand. Incidentally, for the uninitiated, the M1 Garand is not to be confused with the M1 carbine. These are two completely different designs.

Getting back to the M14, it is a very eclectic design. It used concepts from several developmental designs of the time, such as the T20 Garand derivative and the then-new T25. Most of these designs were under evaluation by U.S. Ordnance following World War II. Some of these designs were actually being developed while the war was still raging.

As U.S. involvement in Vietnam was escalating at the same time as the M14 was first being issued to troops, it soon became clear that the M14's wooden stock was going to be a problem for jungle climates. These stocks quickly swelled from moisture and warped. A fiberglass alternative was soon developed, as was an improved fiberglass hand guard.[9] While the recent M14 EBR uses a telescoping stock to help improve handling and reduce length, earlier attempts to provide a short folding stock had been made, with limited success. The primary version used over the years used a rather flimsy folding mechanism that was not known for extreme durability. A similar design was used on the Italian BM59, a 7.62 NATO chambered, M1 Garand based rifle that was used for some time among Italian troops. While the BM59 was a decent rifle, it never became very popular outside of Italy. It was a nicely designed modernization of the Garand but never quite caught on, perhaps for the same reasons that the M14 never became popular outside of the U.S. It used a conventional wooden stock design and maybe just seemed to represent an earlier generation of weapon design.

The M14 proved to be a highly effective sniper rifle and was used as such starting in the mid–1960s. It served for a time during the mid- to late eighties as the standard U.S. Army sniper rifle.[10] An original sniper scope

This M14 is fitted with a fiberglass stock, mount and variable power scope. The scope allows the shooter to take full advantage of the M14's long effective range. The fiberglass M14 stock was developed during the Vietnam era, as wood stocks quickly swelled and warped from the severe humidity of the jungle. (MCS V. J. Street)

was designed for troops in Vietnam and was known as the ART (Auto Ranging Telescope). The designer was an officer named Leatherwood.[11] These Leatherwood scopes used two horizontal stadia wires to bracket the target to provide a range value and adjusted the elevation accordingly. The scopes were very successful and are still in use today. Commercial versions have been offered for some years.

While the M14 only served for a few years as standard service rifle, variants serve in limited roles even today. Other sniper variations of the M14 have been developed since the 1970s. One of the more well-known versions is the M25 sniper variant. The M25 differed from the standard M14 in that it used a synthetic stock and a Harris lightweight bipod. Several modified parts were also used to improve performance. An improved scope mount was also standard, as was a match grade barrel. The M25 was set up for semi-auto fire only but could be restored to select-fire mode if needed.

In the early 1990s, the U.S. Marine Corps began using a special version of the M14 known as the DMR (Designated Marksman Rifle). These rifles also use adjustable synthetic stocks, Harris bipods and match grade barrels.

The most recent versions of the M14 are models designed to bring the rifle in line with other modern service weapons. The M14 EBR (Enhanced Battle Rifle) was first designed in 2003 and has become a popular rifle for U.S. special operations units. The EBR is officially known as the Mk14 Mod 0 in U.S. Navy jargon. It is a slightly modified M14 inside of a completely new collapsible stock designed by Sage International, Ltd. The EBR was produced by Smith Enterprises, Inc. Smith has also designed the new M14 SE and Mk14 SEI, highly modified semi-auto versions of the M14 designed to be low-cost alternatives to new semi-auto sniper systems such as the M110 and Mk11.[12] While Smith Enterprises has primarily dealt with the military up to this point, it is likely only a matter of time before civilian versions are made available.

In addition to U.S. M14 variants, the Israelis built a unique M14 version of their own, the M89 bullpup. This model was originally called the M36, but this was unsuccessful due to production issues. The M89 has been made in both iron sight and scope models and provides the Israelis with a compact yet powerful sniper system, which is also available in suppressed form.[13]

Many M14 variants have been designed for suppressor use, and suppressor popularity in combat is gaining, as many advantages exist in their use. This is especially true for snipers, as the use of a suppressor helps to hide the shooter's location. Over the years some M14s have been modified for reliable suppressed operation. Today there are suppressor models that do not require any modification of the rifle other than removal of the flash hider, which is then replaced with the suppressor.

The selector switch of the M14 is somewhat different from that of the

This is the latest popular variation of the M14, the EBR (Enhanced Battle Rifle). While the new stock design adds to the M14's options, it also adds considerable weight to the overall system. The basic empty M14 weighs just under nine pounds, while the EBR weighs more along the lines of eleven to twelve pounds. The weight will increase quickly with accessories. This makes for tough patrols in thin air. (Snr. Airman Grovert Fuentes-Contreras)

FAL or G3, both of which used selectors that were very similar in design and operation. In contrast, the M14 used a selector that was connected to a long bar that was part of the connector assembly. This long bar was in contact with the sear when set for automatic fire and was activated by a projection on the operating rod after the bolt was locked and the rod was finishing its forward travel. This design ensured safe automatic operation and was simple yet clever at the same time. Activating only when the carrier was finishing its return motion ensured that the bolt was fully locked and there would be no issues with premature ignition. If a locked breech rifle fires without the bolt fully locked, the shooter is usually injured or worse.

The gas system of the M14 was self-regulating, as previously mentioned, and there was no method of adjustment. There was no need for any adjustment, as a fouled rifle would eventually force the piston free unless completely seized, which was highly unlikely. The piston would move to close off the barrel gas port after it had enough energy to cycle the rifle. One thing that should be noted: the M14 used a stainless steel gas piston machined with very close tolerances and no oil is recommended for reliable operation. Chinese models do not necessarily use these same pistons and regular steel has different characteristics. Some oil may be needed on these. This is not a recommendation and users should contact the manufacturer for proper instructions on their particular rifles. Some users have reported using oil on even the stainless piston with no problems. However, the U.S. Army M14 field manual instructs that the piston and cylinder be kept dry. The oil mixes with the powder fouling and creates a gummy compound, which may create pressure issues and cause sluggish rifle operation.

Proper lubrication is needed for any firearm and the M14 has several points of wear that require attention and may need either oil or grease depending on the location. One good rule of thumb is to lube any area that shows wear from normal operation. On the M14, these areas include the bolt lugs and roller, the operating rod contact points and the receiver contact points. In addition, the trigger group should have a light coat applied to any areas that receive heavy contact, like the hammer nose and the sear engagement area. Grease is popular for areas that receive heavy friction and heat, like the bolt lugs and roller channel. While several types of grease and oil have been developed over the years, high-temp bearing grease has always proven a good choice for grease and any good synthetic gun oil will do a fine job, especially one that adds a low-friction compound like Teflon. Oils intended for use in extreme cold should be specifically developed for such climates. These maintenance rules apply for all firearms, not just the M14. Some weapons have specific maintenance requirements and the user should be familiar with them and should always read the particular weapon's operator manual and contact the manufacturer if any questions still exist.

The U.S. had pretty much pushed the M14 out of the picture until operations began in Iraq and Afghanistan. Since that time, use of the M14 has doubled and in some cases its use is probably needed as a primary service rifle, as it is the only standard rifle we have that possesses the range needed to engage the enemy at the distances from which they have quickly learned to operate.

The M14 has proven to be a highly versatile and effective service rifle that can deal out accurate fire to a great range, far beyond what the M16 series can deliver. The need for this has been made all too clear in Afghanistan, with the terrorist groups quickly learning the maximum effective range of M16/M4 fire and adjusting their attack ranges accordingly.

As those familiar with the M14 already know, the design of the rifle was hardly original, even at the time of its introduction. It was widely recognized to be nothing more than a product improved M1 Garand. In actuality, this is partly true. It was far more involved than that, however. The M14's gas action was quite different from that of the Garand, though there were obvious similarities between the two rifles. The trigger group was almost identical, as was the bolt and much of the operating handle design. In addition, the sights of both rifles were nearly identical.

The beauty of the M1 Garand design and, by default, much of the M14's design as well was its straightforward and simple layout. This same simplicity is also, partly, what made the AK-47 such a successful design.

By the end of World War II, only two countries had adopted a semi-automatic rifle as standard issue in numbers that can be considered significant. These were the United States and the U.S.S.R. The Germans had several designs that saw some limited use during the war, like the StG 44 and the FG 42, but none were ever used in large numbers. By the time these designs had been readied for service, German production capability had been severely hindered due to the Allied bombing campaign. Of the two standard semi-automatic rifles, the M1 Garand was by far the more field worthy. The Soviet Tokarev SVT 38 and the improved SVT 40 were in need of some major improvements before they could reach this level of performance.[14] Their recoil tended to batter these rifles severely during use.

While the M1 used a reliable gas system, its design was far from perfect and it generated rather abusive stress on the piston and operating rod mechanism. The M14, by comparison, had been developed to provide a gentler cycling of the action. This new gas system design, however, tended to isolate the operating gases in a confined area, which would get quite dirty after extended firing periods. Proper cleaning of this gas system involves a fair amount of elbow grease. The nice thing about the design of the M14 is that this gas system allows the rest of the rifle to stay as clean as a whistle. The M14 can fire hundreds of rounds and the breech area will often look as if it were just cleaned. Crucial in the M14's maintenance regimen is the proper cleaning of the gas system. The operator must always remember to leave the gas system of the M14 dry after removing all the carbon buildup. This can be done with any standard cleaning solvent.

Cleaning the M14 is no different from cleaning any other service rifle. Proper cleaning of the barrel and chamber area along with the bolt and other components is all that is required provided proper lubricants are applied in the appropriate location after cleaning.

Field stripping the M14 is a relatively simple procedure. As usual, the first step is to clear the weapon by removing the magazine. As the M14 uses a paddle type magazine release, this technique should be familiar by this time. The paddle is pushed forward and the magazine is pivoted forward from the bottom and pulled free from its well. After removing the magazine, pull back on the operating rod handle and visually inspect the chamber and magazine well to ensure that they are clear. Slowly return the operating rod to its forward position and place the safety in the on position by pulling inward toward the trigger. This last operation cannot be done unless the hammer is cocked.

Disassembly of the gas system is very simple. To remove the gas piston, simply unscrew the gas plug from the front with a proper fitting wrench and allow the gas piston to slide forward. If it is stuck in place, a proper size tool can later push it forward when the weapon is further disassembled.

To separate the trigger group from the weapon, pull the rear of the trigger guard back and down to swing it free from its locking latch. With the trigger guard opened, the trigger group can now be pulled straight down and free from its alignment grooves within the receiver. At this point the barrel receiver group can be pulled up and free from the stock. Keep in mind, there is a retaining lip at the front of the stock. After lifting up the rear of the receiver, push the barrel receiver group forward and clear of the retaining lip in order to separate the stock.

The next step only applies to those M14s with the full-auto connector assembly installed. Press in and turn selector knob until the "A" (yes, for "automatic") faces up. Push the entire connector forward far enough to clear the connector lock pin. Swing the connector arm down until it can be lifted clear of its rear retention point. This will require some tilting and wiggling on the connector rod before lifting it free.

After removal of the connector, remove tension on the recoil spring and slide the connector pin out. Next slowly release the recoil spring and guide slightly to allow it to clear its channel. At this point, pull back on the operating rod handle until it can be wiggled out from its retention points in the receiver. This will also require a bit of tipping and tilting. The bolt can then be removed by doing the same. Make sure not to force the bolt or operating rod while doing this, as they could be damaged.

If needed, the flash hider can be unscrewed with a special wrench. The gas cylinder lock can also be unscrewed from the front and the gas cylinder removed if desired. If the piston did not come free when the gas plug was removed, it can now be pushed forward with a small wooden dowel or other proper fitting tool at this point.

This is all that is necessary or recommended for disassembly of the M14 service rifle for maintenance purposes. Reassembly is simply the reverse, as is typical. When reattaching the gas cylinder lock, remember to back the cylinder lock off one turn from snug until the holes are aligned to allow for piston travel.

The M14 has received more attention in recent years, due to the war in Afghanistan, than it ever received during its short period as the standard U.S. infantry rifle. The M14 is a long lived rifle, due much to its excellent quality of construction and a well-developed design that was the result of many years of teething troubles.

5

SIG SG542

While the rifles previously covered were the predominant models used since World War II, they were not the only 7.62 NATO rifles developed since the close of the war. The standard Swiss service rifle for much of the Cold War era had been the SIG StGw 57, first adopted in 1957.[1] This rifle was known for its excellent quality of design and construction, as were all weapons developed by SIG over the years.

The StGw 57 was very similar in operation to the HK G3 in that it used a roller-locking delayed-blowback system. The parent design of the StGw 57 was an earlier Swiss model, the AM55.[2] This rifle used a great deal of the technology developed by the Germans during the war. The production model of the StGw 57 looked more like a light machine gun than an assault rifle. It used a straight line barrel/stock layout and a folding bipod was standard, as was synthetic stock material. This rifle was chambered for the 7.5×55mm Swiss cartridge and never sold well beyond a few numbers to developing nations (mostly in South America). It was an excellent rifle, but its cost was prohibitive.

A modified version was designed to help increase export sales. This rifle was known as the SG510-4. It was chambered for the 7.62mm NATO round to help with sales potential. SIG went a step further to make the SG510-4 more appealing to foreign armies. The stock and forearm were made of wood (which seems like a step backwards) and had a more conventional shape. The 510-4 still carried the normal pistol grip and raised sights, but they were a shorter pattern than the folding sights of the StGw 57 (aka 510-0).[3]

There was also a 510-2, which was a lighter version of the standard 510-1. The 510-1 was similar to the 510-0 but was a commercial rifle chambered as requested. The 510-3 was never to see heavy production and was chambered for the 7.62×39mm Soviet caliber. This was a poor idea from the start given the cost of the SIG when compared to that of the AK-47, which utilized the same cartridge.

While the SIG 510 served the Swiss military well for years, its cost all but ensured it would never be a popular candidate for export sales. As a result, SIG had continued to develop new rifles to keep pace with other manufacturers. The first of these designs was the SG530. This rifle used a gas-operated roller-locking system that proved complicated and also expensive, although it was supposed to be a cheaper alternative to the 510 series.[4] As a result, SIG went back to the drawing board and came up with a new, AK-47 based gas-operated rifle, the SG540. Here is where the legend was born. This would prove to be an export success and would eventually lead to the current SG550 series, the new standard for the Swiss military. The SIG SG550 series are often considered to be the world's best military rifles currently offered.

While design of the unsuccessful SG530 began in the '60s, the 540 series was quicker to get up and running and was ready for production by the very early 1970s. Licensed production took place in the late 1970s at the French arms manufacturing plant known as MANURHIN (Manufacture de Machines du Haut-Rhin). The reason usually accepted for this foreign production facility is Swiss export sales concerning belligerent nations. Setting up production facilities in another state allowed SIG to bypass these restrictions. Since the 540 series was being produced in France, sales to African nations became commonplace.[5] It also sold well in South America. While European production continued into the late 1980s, current production of the SG540 series today takes place in Chile at the FAMAE (Fábricas y Maestranzas del Ejército) national arms plant.[6]

Three basic variations of the SG540 series were made over the years and this tradition continues at the FAMAE plant today. The FAMAE 540-1 is the basic 5.56mm rifle and is the equivalent of the SG540. There's also a carbine version of the rifle known as the 543-1 that matches the original SIG SG543. The final version is the primary focus of this chapter, the 542-1. This is the current version of the original SIG SG542. This final rifle is not a 5.56mm but is chambered for the 7.62mm NATO cartridge.

In actuality, the SG540 series does not differ much from the later 550 series. The operating system was left almost unaltered, but a plastic magazine and a new buttstock design were added. The sights were altered only slightly. Incidentally, the 540 series rifles were and are (at FAMAE) offered with either a fixed plastic or folding metal stock. The 550 series differed in having one standard folding plastic stock design.

The FAMAE 542–1 with folding stock. While not often seen, this is one of the world's best 7.62mm NATO rifles in terms of reliability and is fairly light at 8.4 pounds empty. As with many modern rifles, a range of accessories can be fitted to the gun. These will change the handling characteristics considerably. (U.S. Army)

The SG542 and 542–1 are, oddly, not encountered in large numbers worldwide, though they are standard issue in several Latin American military and police units. This is a shame, as they are quite probably the best overall 7.62mm NATO rifles currently made, though a new Swiss 7.62mm rifle (more closely related to the SG550) has been recently introduced. Even its heavier folding stock variation, the 542–1, weighs less than eight and one-half pounds empty. It has a barrel length of roughly 18 inches, which is an excellent choice for this caliber, as it allows for good handling and portability while still maintaining decent ballistic potential. A 7.62mm NATO rifle with a barrel length of less than 18 inches is still burning too much powder as the bullet leaves the barrel, and the flash and blast become excessive. Velocity also suffers considerably with the shorter barrels like those found on the 7.62mm Galil SAR (usually over 200fps drop when compared to the traditional 22").

The standard magazine for the 542–1 is made from steel and holds twenty rounds, which is the most common capacity for a rifle in this caliber. This is likely due to ammunition weight. Occasionally twenty-five- and thirty-round magazines are seen in this caliber for rifles like the 7.62 Galil and FAL, but while the Galil twenty-five-round magazines work reliably, the longer thirty-round FAL magazines have been known to have an issue regarding feed reliability. Part of this issue may have been due to the practice of stretching magazine springs to help with tension issues.[7] This was done at times in U.K. military units that used both the FAL rifle (L1A1) and the Bren LMG (L4A1). These two weapons use interchangeable magazines. The problem was that the thirty-round magazine of the L4A1 was designed to feed from the inverted position while the FAL had to fight gravity to feed rounds into the chamber. At any rate, the twenty-round 7.62mm magazine capacity is generally regarded as standard for this caliber and most battle rifles.

The 542–1 comes standard with a hard chromed chamber and this is fitted with a closed end flash hider. The sights are fully adjustable and the front sight post is well protected by two large vertical ears. The rear sight is a rotary drum pattern very similar to that used on the HK G3. The current Chilean rifles differ from the original Swiss models in that optic rails are offered as a standard fitting atop the receiver. A removable three-round burst fitting is available on all models and this was a carryover from the original SIG versions. The SG550 series use a three-round burst setting as a standard, permanent fitting. A lightweight folding bipod is an option and a good choice for the 7.62 model to allow the user a stable firing platform for long-range shots.

The action of the SG540 series is very similar to that of the AK-47 in layout, the primary difference being the 540's adjustable gas regulator, which has one position for normal firing and one for conditions of severe fouling. The second setting allows for a larger volume of gas to be delivered to ensure reliable functioning. A third position is meant for firing rifle grenades and closes off the gas system. The AK, in comparison, uses a one-piece gas piston and bolt carrier assembly where the SG540 series use a piston attached to the bolt carrier and held in place by the cocking handle. All three are separated during the disassembly procedure. The recoil spring is also located in a different position from that of the AK-47. On the AK the spring is positioned after the bolt carrier, where on the 540 series it is a coil design located around the piston itself, located ahead of the bolt carrier. This type of spring layout has often created problems regarding heat treatment of the spring due to its proximity to a hot barrel. This issue, however, has usually only applied to machine guns that fire heavily in full-auto.

The bolt carrier design is pure AK in lock-up pattern, and the SIG has built a similar reputation for reliability. SIGs have an added advantage of being among the highest-quality rifles ever built. Unfortunately, this adds considerably to the price of the rifle. This cost is in spite of SIG making every effort to use the latest in manufacturing technology, such as stamped receivers rather than the forged/milled variety. Quite often, quality just costs more. Standard SIG fittings are also not usually seen on the AKs. The rotary aperture sights, folding bipods, and three-round burst options are all SIG only features. However, the basic AK guts remain.

Sadly, no import versions are available for the commercial market here in the U.S. For that matter, very few of the SIG 550 semi-auto models were ever imported before the 1989 import ban on many semi-automatic rifles. If one is found on the used gun market, its price will be extreme, to say the least. To fill the demand for SIGs here in the U.S., American made variations have been introduced recently. However, only one is offered in 7.62 NATO, the SIG 716.

The 716 is not a semi-auto version of the SG540 or SG550 series. It is a new design entirely and is a short-stroke, gas piston AR type rifle. Due to demand for the more traditional SG540/550 long-stroke AK based action, SIG recently introduced a 5.56mm only copy, the SIG 551-A1. It is a shame that the SG550 series never had a 7.62mm NATO variation, but time will tell if a 7.62mm version will ever be offered in semi-automatic commercial form. If such an option were offered, it would likely find a ready market in the U.S. With renewed interest in the gun control debate due to several mass shootings in the U.S., many firearms manufacturers may be holding off spending money on new weapons development for the civilian market.

SIG has recently developed a new 7.62 NATO rifle that is more faithful to the 550 design, in the form of the SAPR (Swiss Arms Precision Rifle), also known as the SG751. While it is not available in the U.S., this new design follows the line of the SG550 series much more closely than the 716. The new SG751 is basically the SG542 that has gone through an updating process similar to the one the smaller SG540 went through in order to become the current SG550 series.

The new SG751 very much resembles the models in the 550 series. The 751 uses the same plastic skeletonized folding pattern stock. Using this stock design helps both to save weight and to create a more comfortable cheek weld than is allowed with the folding tubular metal stock of the earlier SG542.

The new SG751, while primarily being an enlarged 7.62mm NATO variation of the SG550, has a few new additions. For starters, the lower receiver is no longer stamped steel as on the original SG550/551. A new lightweight aluminum alloy lower receiver was developed for the ultra-short SG553 5.56mm carbine several years ago. Using this as a starting point, a similar type receiver was developed for the new larger 7.62mm weapon. The advantage here is in a considerable savings in overall weight versus what would have been seen with a stamped steel lower receiver. The new full-length barreled SG751 is listed by SIG as having a weight that even the FAL

50–63 carbine and FN SCAR-H cannot beat, though the shorter barrel SCAR-H models are about the same weight as the SIG.

While the SG751 is still new, it is not much more than the SG550 in larger form. This is aside from the alloy lower receiver, which receives little stress from the recoil forces created from firing. Many of the accessory options are quite useful and well thought out. This is not surprising, as the smaller 550 series has been around since 1983.

In addition to the original pattern forend, the popular quad rail pattern is offered. The original 550 pattern folding stock can be replaced with a length of pull adjustable stock similar in looks to the original stock. As a final option, an M4 pattern telescoping stock is available. For use aboard helicopters, a brass catching attachment is offered. Several different grip options are offered, as is the original 550 style grip. One optional grip offers finger grooves for enhanced grip and comfort. Another grip is fitted with a palm shelf to prevent slipping when hands are sweaty.

The same 550 pattern cheek piece can be fitted to the stock, though this is primarily needed only when high-profile scope rings are used, as a rail equipped receiver is standard on the 751. This is an option for the 550 series, as both "flat top" and iron sight versions are available on the smaller 5.56mm models. The folding sight equipped flat top models offer the advantage of being able to utilize any modern optics without the need for the cheek rest attachment. This is an improvement, as the additional components tend to reduce the handling quality of any weapon, in addition to adding unnecessary weight.

With the shortcomings of the 5.56mm and the realization that the 7.62mm NATO is far from washed up, the birth of this new SIG has taken place at an appropriate moment. It is, at heart, the big brother of what is, more often than not, considered the finest assault rifle in the world.

The 1970s was a major turning point for SIG. The company had a long history of making quality products, but with the coming of the 1970s SIG was to see a series of successes that gave them the momentum to move forward to the position within the firearms industry that they enjoy today.

While the SG540 series that arrived in the early part of that decade certainly proved to be the successful design that SIG was looking for, it was not as big a move for SIG as the 550 series would prove to be. SIG's biggest move during this time, however, was its development of a new service pistol intended to replace SIG's excellent but extremely expensive P210 9mm. The result was the very modern P220 series. This design would soon give SIG the reputation of being the makers of the world's finest combat pistol. In fact, the P220 was made in Germany by the firm of J. P. Sauer & Sohn. Once again, this manufacturing arrangement allowed SIG to export the weapon to countries with which they could not have otherwise done business.

The twenty-round magazine has also been redesigned and is now made from polymer rather than steel. Unlike the 5.56mm models of the SG550 series, the SG751 comes in a flat top, rail only receiver pattern. There is no separate iron sight model, though a detachable rear aperture sight is offered, as is a folding front sight. The barrel can be changed out and several barrel lengths are available. Lengths of 14" (SB), 18", and 20" (LB) are standard. Twenty-two inch may be offered as well. All barrel options are cold hammer forged to increase service life. Also available is a quick detachable suppressor and a quad rail forend, which replaces the standard version. The standard pattern forend is quite similar to that of the SG550 series. The lower receiver is offered in either semi-auto or select-fire variations. The overall weight of the new SG751 is roughly eight pounds without magazine. The longer barrel models add to this weight. This reduction in weight is due primarily to the new stock and the use of aluminum in the construction of the lower receiver. How well the alloy receiver will hold up over time remains to be seen. Aluminum generally doesn't offer the same durability as steel where firearms are concerned.

Much like the SG542 that came before it, the SG751 has everything one could ask for in a combat rifle. It has the range and power delivered by its 7.62mm NATO chambering. It offers the reliability of an AK-47 derived action. In addition, it also has the nice extras often lacking on the standard AK. Features like the easy to manipulate safety/selector and last round bolt catch are not found on the run-of-the-mill AK-47. SIG also traditionally offers a quality of construction seldom seen on other small arms. Combined, these features have given SIG rifles the reputation they enjoy today, that of being what many consider the finest overall military rifles ever built.

Although there is not a great deal of difference between the SG542–1 and the new SG751 internally, aside from the lower receiver change, the new SG751 does bring with it some minor changes similar to those found on the SG550 series. This in essence gives SIG the newest and perhaps best 7.62mm NATO battle rifle currently offered. It is unfortunate that SIG had not introduced the rifle before the economic downturn, as this would have no doubt helped sales. Because of their high price, SIG rifles in general do not tend to produce large sales orders, but perhaps the renewed interest in the larger caliber rifles due to recent experiences in the Middle East will influence the future of the SG751. There is even potential for possible large sales orders from select law enforcement agencies or special operations units around the world. While it is unlikely, due to current economic conditions worldwide, the SG751 may even see adoption as a standard infantry rifle by some nation that does not have an indigenous design. SIG's reputation has helped in this regard over the years. As long as the production quality that all SIG rifles have long been known for does not waiver or change, the SG751 certainly has a great deal of potential to succeed.

Although the 5.56mm versions of the SG540 series have traditionally been more successful, recently realized shortcomings of the 5.56mm NATO cartridge have renewed interest in 7.62mm rifles worldwide. This bodes well for the future of the FAMAE 542-1 as well as the new SIG.

The 542-1 comes in two basic models, one with a fixed plastic stock and an empty weight of just over eight pounds and one with a right side folding metal stock with an empty weight of around eight and one-half pounds. The barrel length on both models is roughly eighteen and one-quarter inches. The cyclic rate of fire is roughly 650 rounds per minute.[8] The upper receiver is equipped with an M1913 Picatinny rail for optical mounting. The gas piston lies above the barrel, in keeping with the AK lineage. The upper and lower receivers are both made of steel. In contrast to the standard AK-47 pattern, the firing pin has a return spring to help prevent "slam fire" incidents. A slam fire is the firing of a chambered cartridge due to nothing more than the forward momentum of the firing pin. This occurs when the bolt carrier is driven forward under force of the return spring. It usually only occurs with overly sensitive primers, but it does happen on rare occasion. Some later AK type rifles have added safety measures such as firing-pin springs to eliminate this type of potentially lethal accidental discharge. This type of accidental discharge is not unique to the AK-47. It can potentially occur with any rifle design that has this type of unblocked firing pin (M16 included). This seldom occurs due to the limited mass of the pin itself, but some of the more sophisticated designs have some form of safety addition to prevent this. For example, the South African R4 AK variant uses a synthetic bushing to hold the pin to the rear during recoil. This bushing compresses during firing and returns the pin once the hammer nose is clear of the bolt's rear face.[9]

The magazine release of the 542-1 is the conventional paddle type, and the magazine must have its forward notch engaged and then be tilted up into its locked position as with other paddle type magazine wells.

The original SG540 rifle was quite widely adopted for some time, though most purchases were from developing nations and those with ties to France, as France was far quicker to export the rifle than was Switzerland. The SG540 was used by several French military units for a time as standard issue due to delays in developing the FAMAS bullpup. During this time, the SG540 apparently made quite an impression, as reports are France is unhappy with the performance of the FAMAS and the current SG550 series is in the running as a possible replacement rifle.[10]

The SG540 and SG550 series rifles have earned an awesome reputation since first introduced around 1972.[11] What else should be expected when combining the extreme reliability of the AK-47 action with Swiss precision and attention to detail? While the SIG 540/550/751 series look far more sophisticated than the AK-47, there are only slight internal differences between the two designs.

The gas system on the AK is fixed, where on the SIG it is an adjustable unit with settings for varying degrees of fouling or suppressed/unsuppressed use depending on model. While the takedown procedures to be covered apply to the current standard SG550 series, the FAMAE 540 series differs only slightly due to minor improvements in the piston/gas valve area for the newer 550 series. These improvements were designed to make for softer recoil and to ease wear on components.[12] Takedown procedures for both are otherwise similar, though the gas systems are somewhat more complicated than the standard AK-47's, due primarily to the adjustable regulator.

The SG540/550 series of rifles, as mentioned, are very much based on the AK-47 design concept. If one were to closely examine the bolt and carrier of the SIG, the relation would be immediately recognizable. Both weapons are long-stroke, gas-operated systems with the gas system located above the barrel. While there is no real advantage or disadvantage to the gas system being located either above or below the barrel, it does allow for a simpler design regarding bottom-fed magazine weapons. SIG may have simply looked at the most effective rifle systems available and opted for the AK-47 based action as the most viable and cost effective replacement for the expensive and failed SG530.

In any case, the takedown procedure for the current SIG rifles series is a simple process. As with all weapons, the first step is the removal of the magazine and then the visual clearing of the chamber. The weapon is then placed on safe, with the hammer left cocked.

To separate the trigger group, press on the rear trigger group pin from both sides to begin removal of the pin, which can then be pulled out to the right side. It is a partially captive pin and its travel will stop. Next swing the trigger group downward and repeat the process with the front trigger group pin. The trigger group can then be pulled free from the receiver. The next step is to press the bolt handle catch and remove the bolt handle. The bolt carrier can then be removed from the rear of the receiver. The handle can be partially reinserted to aid in this procedure if necessary. To remove the bolt from the carrier, push the bolt inward and turn while wiggling it free from the carrier cam recess.

The bottom portion of the forend is removed by pulling it toward the rear and pulling down to separate the back end. Do the same to remove the upper hand guard.

Next separate the gas valve by first pressing the valve catch in and then turning and pulling the valve free. The gas piston and return spring can be pushed forward and pulled free from the front of the cylinder or gas tube. The gas tube can also be pulled out from the front.

The final step is to remove the firing pin and spring. This is done by pressing in the pin from the rear of the bolt and then driving out the retaining pin from the side. The firing pin and spring can then be pulled out from the back face of the bolt. The assembly is the reverse, as is the norm.

The magazine can be stripped if cleaning is needed. Many modern magazines are constructed from polymer and require no lubrication. This is true of the popular Magpul P-Mag used in the M16 series. This magazine is particularly well known for its reliability.[13] Many new U.S issue aluminum M16 magazines use polymer followers and are recommended to be used dry. This lack of lubrication helps prevent the rapid accumulation of grit and greatly increases magazine reliability. In disassembling plastic SIG magazines, begin by pushing in the floor plate retaining pin from the bottom and then sliding the floor plate off to the back. Since the magazines are largely polymer, no lube is needed, as the primary metal component is the magazine spring.

While the SIG gas piston is stainless steel in construction, a light coat of oil is suggested for maintenance purposes.

Cleaning the SIG requires no special attention aside from that usually paid to other gas-operated rifles like the AK-47. The bore and chamber should be scrubbed clean along with the gas tube, gas piston and valve. The trigger group should be lightly lubed along all pivot and contact points. The carrier and bolt should be oiled wherever any wear due to friction is present. The firing pin and spring should also have a very light coat of lube, being careful not to overlubricate this area, as any excess of oil can run down the firing pin and contaminate a chambered round through the primer/case head area. This safety tip also applies to all bolts and firing pins for any rifle.

Though the SIG has always carried a hefty price tag, the rifle is perhaps the ultimate in quality of design and construction, with the end result being one of the most accurate, reliable, well-balanced and effective combat rifles ever designed. The new SG751 will likely continue this tradition for many years.

6

Galil 7.62mm NATO

An event took place in 1948 that many considered unimaginable for over two thousand years: the nation of Israel was established, or re-established to be correct. The new state was immediately set upon by almost every surrounding nation. Since the end of World War II, the Jewish settlers who had been arriving knew that firearms would soon be needed in large numbers and had been secretly smuggling them in from several sources. It soon became clear that the hodgepodge mix of firearms was not logistically feasible and a standardization program was needed with regard to weapons issue. By the 1960s, several weapons had become standard issue for the Israeli military.

The FN FAL 7.62mm NATO rifle was adopted and a locally produced version was soon developed, as was a native submachine gun (SMG) design, the legendary UZI 9mm. The UZI would come to epitomize modern small arms development and is often regarded as the finest SMG of all time. The UZI proved reliable under the most extreme desert sand and dust environments. Unfortunately, the FAL didn't fare nearly as well under these conditions. This was true even when relief cuts were added to the bolt carrier to allow for accumulation of sand. At the same time, the UZI only had an effective combat range of 200m at best, and accuracy even at this distance was marginal. What was needed was reliability with increased effective range in a weapon designed for standard issue.

During the intense and non-stop fighting of the Six-Day War, the shortcomings of the FAL rifle became clear and a replacement was demanded. During this time the Israeli troops had become very impressed by the performance of the AK-47 rifle, which was widely used by Egyptian troops.[1] The call was put out to find a suitable replacement for the FAL, something with reliability equal or superior to that of the AK-47. The Israelis tested several weapons from the U.S. M16 and Stoner model 63 designs, to the AK

The Galil 7.62mm NATO. This is the basic AR variation, which uses a synthetic forend, as opposed to the wooden version present on the ARM model. The rugged folding bipod and carrying handle of the ARM are also missing. The Galil is often considered the "Rolls-Royce" of AK-47s. It has been produced since the early 1980s in this larger caliber and has only recently been dropped from production in Colombia, though Israel quit making it years earlier. (MCC E. A. Clement)

series and even the HK roller-locked system. The reliability of the AK-47 was to set the standard, but they liked the flat shooting 5.56mm cartridge. An Israeli inventor, Yisrael Galili, had put together a 5.56mm rifle based on the AK, using parts from the Stoner model 63 series. The model 63 system was a long-stroke gas-operated design invented by Eugene Stoner as an effort to improve on the performance of the M16 and provide increased modularity in a small arm designed for infantry use. The new Israeli design looked like a winner, and later machined Finnish Valmet m/62 receivers were obtained to put together a more finalized design. This led to the official Israeli adoption of the Galil in 1972. Few of the new rifles were available in time for the Yom Kippur War in 1973.[2] As a result, Israelis fought with the same small arms they had used during the earlier Six-Day War.[3] Once the cease-fire took place and things settled down a bit, the issue of the Galil rifle finally began. While the Galil was Israeli standard for many years, issue of the rifle was never universal. This is because many U.S. M16 rifles were made available at extremely low prices or were given to Israel as part of foreign aid packages.[4] As a result, the Galil has long served side by side with different variations of the U.S. rifle, with the most recent version being the M4A1 carbine. These Colt rifles and the Galil are both being replaced by the new Israeli Tavor bullpup 5.56mm NATO series.[5] The Galil was always liked for its reliability but not for its weight, which varied from roughly eight pounds to over nine and one-half pounds, depending on variation and caliber.

Three basic versions of the Galil were produced, the ARM, AR, and SAR. The ARM and SAR were the two most commonly seen variations over the years. The ARM is the standard or light machine gun (LMG) version if you prefer. This model is equipped with a full-length barrel, a folding carrying handle, and a heavy duty folding bipod. The bipod doubled as a wire cutter and a bottle opener was built into the bipod catch area to prevent troops from using magazine feed lips to open bottles, as this tended to bend the feed lips at times and could lead to feed malfunctions. The wooden forearm used on this model reportedly served better at absorbing heat buildup.[6] This was needed as this variation was likely to see more sustained fire, and the resulting heat coming off the barrel and gas system would likely have melted the plastic forearm seen on the other two variations. Today's improvements in plastics would likely prevent this issue.

The AR model used the same full-length barrel as the ARM, but the bipod and carrying handle were omitted and the forend was changed to plastic. This was meant to be the standard variation, but it appears to have been the least used of the three in service. This may be because it lacked the features of the ARM and did not offer the convenience of the final variation.

The SAR was a short barrel model with the same plastic forend of the AR. This carbine was meant for issue to paratroop units and any other troops who might have need of a shorter, lighter weapon and were willing to sacrifice some effective range and muzzle energy in the trade-off. A folding metal tube stock is standard on all three Galil models. The barrel lengths on the SAR models were just over 13 inches for the 5.56mm and 15.7 inches for the 7.62mm variation.

The Galil was a 5.56mm rifle, whereas our focus has been on the renewed popularity of 7.62mm NATO caliber battle rifles. Luckily, all three Galil models were eventually offered up in the larger caliber as well. In the early 1980s, the 7.62mm NATO Galil was first offered for commercial sale in the U.S. in semi-automatic only, AR and ARM variations. The shorter SAR variation was never offered for commercial sale due to U.S. restrictions on barrels with an overall length of under 16 inches for any rifle. The AR and ARM 5.56mm models were equipped with barrels of just over 18 inches in length while the 7.62mm full length variations used 21" barrels.[7] The barrels of all models were chrome lined for added durability.

A semi-auto only sniper variant was also introduced, shortly after the original 7.62mm models hit the U.S. market. The ARM 7.62mm version has been used by the Israelis but not to the same extent as the 5.56mm models. The 7.62mm ARM model was supposed to provide a support fire capability to Israeli infantry units. This was not an effective SAW, as it suffered from the same affliction as all other rifles adapted for this role. The lack of a quick change barrel on this rifle does not allow for a continuous fire weapon.

The sniper variant used an adjustable wooden folding stock in place of the standard tubular metal folder that was used on the other models. The barrel of this model was a heavier design, with an overall length of 20 inches, and it did not carry the flash hider of the standard models. Instead, it used a muzzle break designed to minimize barrel movement in case a fast follow-up shot was required. The forend of the sniper variant was made of wood but was a different pattern than that used on the other variations. The carrying handle was absent on the sniper variant, and an adjustable folding bipod was attached to the receiver rather than the barrel. This bipod folded forward rather than to the rear as on the other bipod equipped models. The purpose of this change was to allow the bipod to be adjusted by the user and it was also designed to prevent point of impact variance due to pressure from below when the bipod was used, as often happens with rifles using barrel mounted bipods. This sniper model carried considerably more weight at roughly fourteen pounds empty when equipped with its standard scope and quick detach mount. Other Galil models could utilize this mount and it was attached to a fitting on the left side of the receiver.[8]

The 7.62 ARM model weighed just over nine and a half pounds empty, while the SAR version weighed closer to eight and one-quarter pounds with the 7.62mm AR variation weighing about a half-pound more. The bipod on the 5.56mm ARM was of limited value in providing a stable platform for long range shooting, as the 5.56mm 55 grain bullet in use at the time only had an effective range of five hundred yards. The bipod on the 7.62mm ARM model, on the other hand, allowed for a stable rest to allow the user to take full advantage of the 7.62mm's long range ability. However, the bipod did help to stabilize both models when firing full-auto bursts.

The Galil 7.62mm SAR is a fearsome and fast handling battle carbine due to its folding stock and abbreviated barrel. Such a short barrel, however, is not well suited to the 7.62mm NATO cartridge due to the severe muzzle blast and flash characteristics. This has long been a problem whenever any attempt is made to shorten the barrel of a weapon designed to use a rifle cartridge. The problem is simply magnified due to the power of this round. The 7.62 SAR does make for a maneuverable carbine in tight spaces, making a very useful urban combat weapon that still brings with it a great deal of hitting power due to its caliber.

The primary difference between the operation of the Galil series and any standard AK-47 is in the use of the safety. The AK safety was never known for being positioned in a convenient location. The Israelis helped alleviate this shortcoming by adding an easier to reach thumb lever on the left side of the pistol grip. The standard right side lever was still there; it was just mechanically connected to the thumb safety through the receiver. The Galil also offered a vertical cocking handle that was easily manipulated by either left- or right-handed users. One final improvement many Galils have is a firing-pin spring to prevent "slam-fires," though the original models did not have this feature. The cyclic rate of fire was a comfortable 650 rpm.

An improved version of the Galil sniper rifle was introduced over ten years ago. The SR-99 was still the same basic Galil sniper rifle but used plastic for a newly designed folding stock and forend. The new stock design also contained an adjustable monopod, which allowed for improved stability. An improved pistol grip with a bottom hilt (palm shelf) was added to prevent the hand from slipping off under stressful firing conditions. In addition to these changes, a band was fitted the length of and just above the barrel to break up the heat rising from the barrel, which could cause visual distortion when using the optics. This was a useful feature, as these heat vapors often negatively affect the shooter's aim. The primary improvement was in weight reduction. The new SR-99 weighed about one and one-half pounds less than the original the Galil sniper rifle.[9]

The most recent sniper model uses a Harris type lightweight folding bipod attached to the forend and yet another new type of folding stock has been added, though the monopod is no longer present. The overall weight is similar to that of the original sniper model.

The Galil uses a protected front sight, which is adjustable for both elevation zero and windage. The rear sight is a flip type L-shaped aperture with 300 and 500 meter range settings. Folding radioactive night sights are standard as well. The advantage of the Galil's sights over those of the conventional AK pattern is a longer sight radius due to the rear sight being receiver mounted and the use of an aperture rather than a notch for the rear sight, which allows for faster target acquisition. The sight radius wasn't that much of an improvement, as the front sight of the Galil was also moved to the rear and was located at the front of the gas tube. On the standard AK-47, the front sight is mounted to a raised bracket located at the front of the barrel. The use of the aperture is a big improvement, however.

While the Galil was never as widely used by Israeli troops as they may have intended, it sold fairly well overseas and has proven popular in Latin America in particular. It is so popular in that region of the world that Colombia (with help) developed a new rifle system based on the original. The new design is called the Galil ACE.[10] This new rifle is available in 5.56mm, 7.62mm NATO, and 7.62×39mm Soviet. Several barrel lengths are offered in each caliber as well as a compact 5.56mm variation, which differs slightly from the standard 5.56 versions. The 7.62mm NATO version will be covered in a separate chapter.

Despite its proven performance, the 7.62mm Galil models are no longer offered by Israeli Weapon Industries, Ltd., and only certain 5.56mm models are still listed along with the most recent 7.62mm sniper rifle variation.[11] This is unfortunate, as the 7.62mm Galil series were among the most rugged, well-made and well-designed service rifles of all time.

For civilians interested in owning one of these rifles, a used semi-automatic Galil will likely prove prohibitively expensive, as they have not been imported since 1989 and owners of such a well-made and durable design are rarely willing to part with it. Time will tell if the newer ACE derivative can match the Galil's record for reliability.

While the SIG SG540/550 series kept the heart of the AK-47 action, they use more developed and adjustable gas systems among other extras and have a look all their own. The Israeli Galil, however, is unmistakably AK. The receiver is nearly identical to the Finnish AK variant, the m/62. The Galil just has a great many extras added to it. These are not present on standard AKs.

Semi-auto civilian Galils differ from the military variations in that they lack the auto sear and its related components. Corresponding alterations are made to internal dimensions to prevent installation of these parts as is usual for civilian variations of military rifles.

The safety system is somewhat different on the semi-auto only Galil. On the standard selective-fire models, the safety is on when the left side thumb lever is fully forward. This is because it is directly connected to the standard AK pattern safety lever located on the right side of the receiver. As the right side lever is pushed down, the left side lever moves to the rear. The left side lever first moves to the auto position then to the semi-auto position. The arrangement is actually one done for mechanical simplicity and not for tactical advantage. This is true on several levels. First, the movement of the left side safety is the reverse of what is natural for the shooting hand, as most safety levers move forward or downward to the firing position. This reversal of the lever's movement makes the Galil's left side safety slower and more awkward to operate than many other rifles.

The second issue is that the semi-auto position is the second position the selector must move to rather than the first. Generally speaking, troops who are forced to carry a limited supply of ammunitions try their best to conserve ammo and use automatic fire only when absolutely necessary. This doctrine is stressed to Israeli troops more so than many others. The Galil's safety makes the move from "safe" to "fire" a two-step process, which is slower but, more important, loud as well. There is a distinct click with each step of the selector lever. This is a tactical shortcoming of the Galil (and the AK for that matter). In practice, it is not a severe shortcoming, as if true stealth is required and shooting is expected safety levers are often moved to fire positions prior to contact with enemy troops. The shooter's finger is kept outside of the trigger guard and off the trigger until contact is made. This is not often recommended during training, but troops quickly learn to take every advantage available to them in order to increase their chances of survival.

Regarding the safety lever of the civilian Galil, a more complicated lever hinge system is used. This, however, makes for a more natural motion of moving the safety lever forward to the fire position. In either case, the left side safety/selector of the Galil is an improvement over that of the basic AK pattern rifle.

As the Galil so closely follows the AK line, it would follow that the field stripping procedure is likewise quite similar, which it is. The takedown procedures are the same with all Galil rifles, no matter the caliber or variant. As always, first clear the weapon by removing the magazine. The manner in which this is done depends on whether or not the M16 magazine adapter has been installed. This actually only applies to the 5.56mm caliber models. If this adapter has been installed in the Galil magazine well, then the magazine release is a push button type located on either the left or both sides of the magazine adapter. Later models had an ambidextrous magazine release. If the standard Galil magazine is used, then the typical paddle type release lever is located to the rear of the magazine. Since this chapter is primarily concerned with the 7.62mm Galil, the paddle release is our only option.

The next step of course is to visually check the chamber and magazine well by pulling back the vertically mounted operating handle. This is quite easy to do with the left hand while holding the pistol grip in the right.

After ensuring that the Galil is indeed clear of any rounds, leave the weapon cocked, but do not apply the safety. Next press the receiver cover catch located at the rear of the receiver. Then lift the cover from the rear, slightly tipping it if necessary. Once the cover is freed at the rear of the receiver, pull it to the rear far enough to clear its engagement lip at the front of the receiver. After removing the receiver cover, press the catch in until it clears its retainer, move it slightly to either side and slowly allow the recoil spring to relieve its tension. The recoil spring and guide can then be pulled clear of the receiver and out of its recess in the bolt carrier.

Next pull back the bolt carrier until it can be lifted clear of the receiver. Again, some manipulation may be necessary here. The bolt can then be pulled out and turned free of the carrier.

The final step in disassembly is to remove the gas tube by pulling it back and up, free of its recess.

The standard steel magazine for the Galil can be stripped in the usual fashion, by sliding off the floor plate and removing the guts of the magazine. As the Galil's standard magazine followers are constructed of metal, a very light coat of lube is a good idea.

The Galil operator's manual does not recommend any further disassembly.

Cleaning procedures for the Galil are almost identical to those for any other AK variant. The usual treatment for an AK gas system is to let it run dry with no oil. The Galil operator's manual actually says to lube this area. Other than this one discrepancy, maintenance techniques for either design are the same. The gas piston of the Galil was hard chromed, so the recommendation to lube this area is unusual.

Make sure the barrel and chamber are scrubbed clean with the appropriate bore or chamber brush, respectively. Ensure the gas piston is thoroughly cleaned along with the gas tube or cylinder, and make sure that the gas block port is clean and free of any blockage due to fouling. Pipe cleaners dipped in cleaning solvent are excellent for this purpose. Make sure the bolt face is clean and ensure that the extractor and ejector areas are not overlooked. The locking lug recesses in the receiver should be cleaned and lubed along with the cam and locking lugs located on the bolt. The carrier should be lubed at any areas showing friction wear. The recoil spring should always have a light coat of oil as well. As usual, the trigger group should also be cleaned so that it is free of any grit and then lightly lubed at all friction and pivot points.

These basic weapon cleaning procedures will ensure any AK functions reliably under all conditions, although AKs usually function reliably even if these maintenance measures are not religiously followed.

On the Galil and any other weapon fitted with night sights that contain radioactive elements, a nylon or other non-abrasive brush should be used to clean the faces of these elements, as harder metal brushes will eventually scratch the containers. Alcohol dipped cotton swabs are good for cleaning night sight inserts, unless the manufacturer recommends otherwise.

The normal day sights should also always be kept clean, and this is especially true for the aperture of the rear sight. This is one area where an obstruction of any kind makes hitting the target a crapshoot.

The folding stock assembly of the Galil is a rugged component and requires little in the way of cleaning, other than the hinge mechanism itself. Just make sure this area doesn't have any accumulation of grit.

7

FN SCAR-H (Mk17)

While the FN FAL will forever be remembered as the free world's rifle, it only remains in production in Brazil and here in the U.S. in its commercial form. Though the FAL is out of production in the nation of its birth, FN did not get out of the rifle business entirely. While the FAL was busy arming much of the free world, FN had noticed the move towards the small 5.56mm round and made an early but failed attempt at a rifle chambered for this round. The CAL 5.56mm rifle was not a success and never saw full scale production. FN knew that the 5.56mm was likely to become a NATO standard caliber and quickly sought to find a suitable replacement for the CAL. The follow-up FNC rifle was ready for production by 1980 and would eventually become quite successful.[1] This new design borrowed heavily from the AK for its bolt and carrier design, and using a similar long-stroke gas piston, the FNC soon developed a good reputation for reliability and offered a lower production cost than the FAL due to the use of modern stamping technology. The overall weight wasn't reduced by much despite the use of an alloy lower receiver. A common misconception regarding caliber reduction in an existing design is that it will bring with it a reduction in weight. The opposite is actually true if the design was identical other than the smaller bore size. This is because the smaller diameter means extra metal remains in the barrel. While the FNC helped FN stay competitive in the field of modern small caliber rifles, it left FN without a larger caliber battle rifle once the FAL was discontinued. Luckily, FN won a contract in 2004 to develop the Special Operations Forces Combat Assault Rifle or SCAR. This stemmed from a request by the U.S. Special Operations Command (SOCOM) a year earlier. SOCOM wanted a modular weapon system that could be tailored for specific operational requirements.[2]

This meant that the rifle system could be outfitted to handle any situ-

The shortest of the M16 series in service, the Mk18 CQBR (Close Quarter Battle Receiver). This weapon is fitted with a 10.3" barrel, while the M4 series uses a 14.5" barrel. This reduces the effective range of the 5.56mm even further. While this weapon is fine within the parameters for which it was designed, it would be of little use at seven hundred yards. The short rear sight is normal for this model and can be removed if needed. (MC1 J. Foehl)

ation from a field rifle to an urban combat rifle to a weapon meant for close quarter battle (CQB). The SCAR was to be made available in both NATO calibers for increased flexibility and tactical applications. There was also a new grenade launcher developed to round out the weapons family. These became known as the SCAR-L in 5.56mm, also known as the Mk16 Mod 0, the SCAR-H in 7.62mm NATO or Mk17 Mod 0, and the Mk13 Enhanced Grenade Launcher Module (EGLM).[3]

After a long period of development and a later refining process, the first large scale development of the rifle was ready with the SCAR-L being issued to the U.S. 75th Rangers in the spring of 2009. In 2010, it was decided that only the Mk17 or SCAR-H version would be procured in the future along with some conversion kits to allow a switch to 5.56mm if needed.[4] This decision was based on the increased versatility offered by the Mk17 Mod 0, although budget issues may have also been a factor in making this decision. Again, FN has refuted this on their Web site. The Mk13 grenade launcher is also still in the loop. As the Mk17 is the subject of this chapter, this decision suits our needs nicely.

While the smaller, lighter Mk16 uses standard M16 pattern magazines, the SCAR-H was designed to use an original magazine pattern. Maybe there just weren't enough M14 magazines left in inventories to justify matching the magazine well to the M14 pattern. FN may have thought that they could improve on the M14 magazine design, which uses technology dating back to the late 1940s. In either case, the Mk17 uses a standard magazine capacity of twenty rounds, again a typical capacity for 7.62mm NATO caliber rifles. One advantage to this new magazine design is that it can be inserted straight into the magazine well without having to first engage the front lug and then rotate the magazine up into its locked position. This tends to be a much slower process. The magazine catch is similar to that used on the M16 series in both design and location. A big advantage offered by the SCAR is that both its push button magazine catch and its safety lever are ambidextrous. While the safety lever is located in much the same position as that found on the M16 series, the advantage to the Mk17 safety is that the lever does not need to be rotated ninety degrees to change firing modes, which is the case with the M16. With the Mk 17, the lever only needs to be rotated forty-five degrees to switch between the safe position and fire mode. The bolt catch lever is also located in a position that would be familiar to anyone who has used the M16 series, on the upper left side of the receiver above the magazine well. The cocking handle on the Mk17 is far different, however, from that of the M16. The cocking handle is reversible, allowing it to project from either the left or right side depending upon user preference. It is also a reciprocating handle and the operator needs to keep hands clear during firing to avoid possible injury. This is in contrast to the M16 cocking handle, which remains stationary during firing and rests atop the rear of the receiver.

A huge advantage to Mk17 users is the fully adjustable and folding stock. This stock can be adjusted for not only length of pull but height as well. It has an adjustable cheek rest for use with optics and mounts of varying heights. The length of pull adjustments allow for proper fit with users of different height and for proper adjustment when wearing body armor or other tactical gear. The stock of the rifle folds along the right side when a short overall length is required. The projection that holds the stock in place when folding also doubles as a spent case deflector for left-handed users.

The Mk17 uses several sections of rail; one long upper rail and a short side and bottom rails are located in the forend area. The side rails are removable if they are not needed.

The front and rear sights are typical of what one would expect from such a modern design. Both the front and rear sights fold out of the way if optics are to be used. The sights are adjustable for range and windage, and the front can be zeroed for elevation purposes.

A definite advantage that the Mk17 offers over the M16 series, aside from its larger caliber, is in its method of operation. The SCAR uses a short-stroke gas piston design rather than the direct impingement system of the M16 series. The gas system uses a tappet to strike the bolt carrier and send it on its way. This method of operation harkens back to the design of the

M1 carbine from 1941. The SCAR's piston is hard chromed plated and should be left dry during operation, meaning no oil or lubricants of any kind should be used. The rings of the piston should be staggered, as is often recommended on the M16 series for similar purposes. These rings are designed to provide a proper gas seal during operation, and staggering rings helps reduce gas loss. The SCAR's gas regulator is adjustable, and this adjustment is not meant to compensate for various degrees of weapons fouling (as with many adjustable systems) but for when the rifle is to be fired with or without a sound suppressor installed. When a suppressor is to be used, less gas is required to pass through the port to operate the piston, as the pressure doesn't drop off as rapidly as when no muzzle attachment is used.

The barrel of the Mk17 can be changed in a matter of minutes by an operator, but this must not be considered a quick change feature.[5] A specific amount of torque is required, as are unique tools that are not normally included in the typical cleaning and maintenance kit. The amount of force used in mounting the barrel of the Mk17 is designated by the manufacturer and must not be applied as a guess. This is not a sustained fire barrel change feature and a Mk17 barrel change is not going to be performed by the user in the middle of a firefight, as can be done on many modern light machine guns or on the Steyr AUG. However, it does allow users at the operational level the option of selecting the ideal barrel length to suit a particular mission. This is not a feature offered by the M16 series, but then neither is the 7.62mm caliber for that matter.

As a potential replacement for the few remaining M14s, the Mk17 does offer a modern modular 7.62mm NATO rifle that has reduced weight and length when compared to the M14. The Mk17's purpose designed adjustable folding stock vastly increases portability and improves handling in tight quarters. It does all this while providing a clean running gas piston that is easily adjusted for suppressor use. This last feature is a definite plus due to the increase in suppressor use among many combat units within both the military and law-enforcement communities.

In addition to the three standard Mk17 variants, there is a modified Mk20 Sniper Support Rifle (SSR) variant available. The three original variations are the short version, which uses a 13" barrel, the standard version, which uses a 16" barrel, and the long variation, which uses a 20" barrel. The sniper variant differs in that it is fitted with a rigid non-folding stock and has alterations to the barrel receiver area designed to improve accuracy. This model was meant to replace the Mk11 sniper rifle in military use. For those not familiar with the Mk11, it is basically an updated AR-10 type rifle and, along with the M110, both from Knights Armament, Inc., has filled the role

Above: The FN SCAR-H (Mk17) in its standard 16" barrel variation. A short barrel model is also offered along with a long barrel version to suit the mission. This model is proving itself with U.S. special operations units in Afghanistan. An empty weight of roughly eight pounds doesn't hurt. (Naval Special Warfare Command)

Right: This Mk17 is fitted out with a far greater number of toys. A standard length scope is mounted along with a detachable suppressor, infrared laser aiming unit, tactical light and vertical foregrip. This is pretty much every stocking stuffer available to mount on a rifle. (Naval Special Warfare Command)

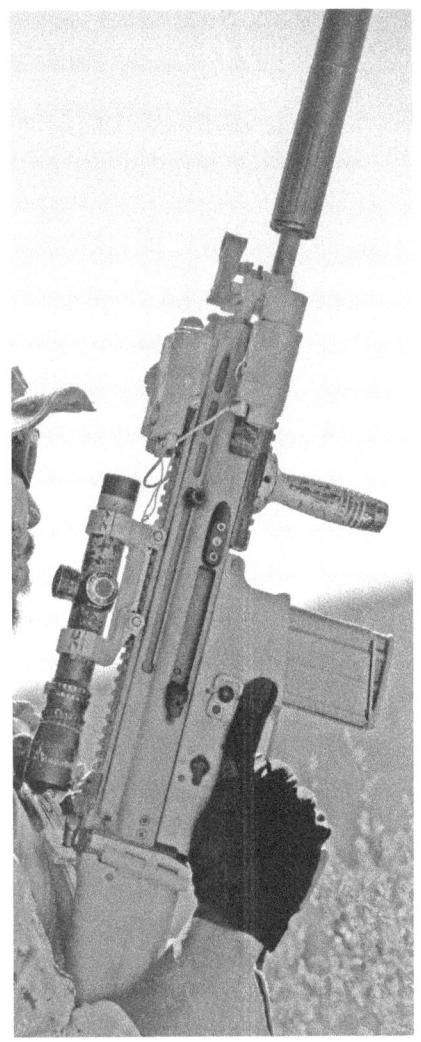

of the semi-automatic sniper system for U.S. troops for several years now.[6] As the Mk11 and M110 are both based on the AR-15 gas impingement design, they suffer the same deficiencies as all other rifles using this method of operation. The Mk20 should improve on this with its more reliable operating system. Due to operational needs and a likely reduced future budget for the military, it is likely that procurement of all SCAR rifles, including the Mk20, will be limited. This of course will depend on whether not it is selected as the new standard carbine to replace the M4. This is unlikely, as the future of the smaller Mk 16 seems to be iffy, though FN has announced that it is not a canceled item with regard to the U.S. military. It is not likely that the U.S. will select the 7.62mm Mk17 as a standard carbine, though this would certainly be a big step up in firepower for the infantry squad.

As for the Mk17 series in terms of weight and portability, the short variation weighs roughly eight pounds empty with a folded length of just under 25½ inches and the length with the stock in its open position varies from 32½ inches collapsed to just over 35 inches in its extended position. The standard 16" barrel variation weighs closer to eight and one-quarter pounds empty with a folded length of just under 28½ inches and a fully extended length of just over 38 inches. Mk20 Mod 0 SSR weighs considerably more at over ten pounds empty, while the overall length is roughly 4 inches longer than the standard model with its stock in the extended position.[7]

For civilian purchase, FN wisely introduced the SCAR in semi-automatic only variations, and these are available in both calibers. The semi-automatic variations are known as the Mk16S in 5.56mm and the Mk17S in 7.62mm. These are purpose built semi-auto only weapons to prevent their conversion to selective-fire capability. Aside from the selector limitations, the Mk16S and Mk17S differ very little from their selective-fire counterparts. This is always a wise move for a manufacturer, as it helps to increase commercial sales potential, as many collectors want the rifle to match its military counterpart as closely as possible. These rifles will likely find a strong market here in the U.S. But this of course will largely depend on future legislative issues regarding firearms ownership, especially ownership of military type semi-automatic rifles that are capable of accepting large capacity magazines.

As a final note, part of the SCAR's attraction is that it just plain looks like the rifle of the future. Time will tell if it can match the performance of the best designs of the past. As there are some top notch rifles that have been made over the years, this will be a difficult task to accomplish. FN, however, has a long history of making excellent products, and so they may very well be able to pull this off.

The FN SCAR, while a relatively new design, is based on proven principles like its short-stroke gas-operating mechanism. The trigger group clearly uses the M16 layout as a starting point, though FN has definitely improved the safety with regard to the user's ability to quickly manipulate the selector lever.

The field stripping procedures are easy to accomplish and no tools are required other than typical cleaning gear, like rod, brush, patches, pipe cleaners, etc.

The first step in field stripping the SCAR is of course to remove the magazine and then retract the bolt to visually verify that the chamber and magazine well are clear. Once the rifle is cleared, place the selector on safe.

The next step in disassembling is to remove the trigger group. This is done by pushing the takedown pin as far as it will go to the right. This is a captive type pin and cannot be removed completely and is located above the front of the magazine well. Next push the entire trigger group forward and then pull downward. This will free the trigger group from its engagement at the rear. With the trigger group removed, the folding buttstock can be pushed straight down and off the rear of the receiver.

Next, pull back the operating handle and at the same time slide the retaining plate, located at the rear of the receiver, downward until it is clear from its engagement. This will release tension on the operating spring and the entire bolt assembly can then be drawn fully to the rear and clear of the receiver. At the same time, the operating handle can be pulled free from the bolt and carrier through the enlarged hole at the rear of the operating handle slot. The remaining portion of the bolt and carrier assembly can be pulled free from the rear of the receiver. The return spring assembly can then be pulled out from the back of the bolt carrier. The final step in disassembling the bolt and carrier group is to push out the firing-pin retaining pin and remove the firing pin and its spring from the rear of the carrier group. Finally, the bolt cam pin can be removed and the bolt can then be pulled forward and clear of the carrier.

To strip the gas system all that is necessary is to first lock the folding front sight assembly in its up position. Next turn the gas regulator knob to its upward (12 o'clock) position so that the stud on the regulator is vertical. Next push in the regulator detent and continue to turn the regulator tab just past its horizontal position (4 o'clock) and pull the regulator out from the front of the gas block. The gas piston can then be pulled out the front of the gas block, using a cleaning rod or other appropriate tool if necessary. The cleaning rod is inserted from the chamber end and then pushed forward enough to allow the piston to protrude from the gas block.

No other disassembly is recommended by the manufacturer and this is as far as the operator should ever have to tear the weapon down for routine maintenance. Reassembly of the SCAR is simply the reverse of the takedown procedure.

Before reassembly, two key issues should be addressed. First, ensure that the gas rings, located in their grooves on the piston head, are staggered. This will help ensure proper gas flow and pressure levels during operation.

The other key maintenance factor is to ensure that the gas portholes on the gas regulator are clean and free of obstruction. Again this is to help ensure proper operation and pressure levels. A purpose made tool for cleaning gas ports is best, but any proper diameter wire porting tool can be used for this.

Cleaning procedure for the SCAR follows many of the same traditions as for other gas-operated weapons. The piston and regulator should be kept dry and free of oil. Whichever solvent is used for cleaning, it should be safe for use on plastics. This applies to many of the modern designs utilizing large amounts of plastic in their construction. FN recommends very light lubrication of the trigger group after cleaning. It is also suggested that the operator not release the hammer when the trigger group is removed. This can cause damage to the trigger unit.

The FN SCAR-H is in use by several select U.S. military units, as mentioned. This is one of the few weapons available at this time to fill the role of a modern lightweight 7.62mm NATO rifle. It appears to be filling this role fairly well so far, with no major complaints from these troops. Minor issues with the folding stock design have been reported, but it is likely that FN has already addressed these concerns, and they will no doubt find a quick cure for any shortcomings in the weapon's durability.

Not only are U.S. special operations units using this new rifle, but recently the British SAS has requested to be issued the SCAR-H in place of the Steyr AUG and Diemaco/Colt Canada C8 carbine series that have been used by the SAS in recent years,[8] though, to be more accurate, the SAS has access to any weapon system that may fit the bill, as with most other special operations units. The SAS has had poor combat experience with the 5.56mm and would apparently like to be re-equipped with a larger caliber weapon that can be more readily relied upon to put an enemy down quickly and at longer ranges than the 5.56mm can provide. Perhaps they felt that the FAL was too heavy, as the SCAR-H does offer a considerable savings in weight over its older brother. It also provides a faster magazine change capability than the FAL, in addition to easier mounting of modern accessories, which have proven very popular in recent years. This may prove to be a trend among special operations units, as they all likely experienced similar combat conditions in recent years regarding the 5.56mm cartridge.

8

HK 417

The Heckler & Koch name has for years been synonymous with the roller-locking, delayed-blowback operating system. Two primary small arms became the company's flagship models, the G3 7.62mm NATO rifle and the MP5 series submachine gun. Other weapons have been developed and successfully marketed over the years. The HK roller-locking system has also been applied to the P9/P9S pistol, the HK 33 and G41 5.56mm NATO rifle, and the HK 21 (7.62mm) and 23 (5.56mm) series belt-fed machine guns, just to name a few. Then suddenly in 1995 HK took a completely different approach and introduced a gas-operated rifle that utilized plastic components, to a great extent.[1] This rifle was, of course, the well-known G36. This was quickly adopted as the standard rifle for the newly reunited German army. At the heart of the G36 lies a self-regulating, short-stroke gas piston operating system. This system was not completely original and owes a considerable amount of its makeup to the Armalite AR-18 design dating back to the early 1960s. The G36 has proven to be a highly reliable weapon system that is capable of functioning even under conditions of severe fouling. Using this reliable system as a starting point, HK undertook the rather daunting task of correcting all of the flaws inherent in the AR-15/M16 series. To a large degree, HK has succeeded. The resulting HK 416 series has been adopted by several military and police organizations, including the U.S. Army's counter-terrorist unit known commonly as Delta Force, which adopted the rifle in 2004.[2] In 2007, it was adopted as the standard service rifle of the Norwegian army. It was also rumored to have been used by several members of Seal Team Six during the raid that resulted in the death of Osama bin Laden.[3] The HK 416 design effectively deals with the well-known issue of "carrier-tilt" that is sometimes a problem for gas piston AR type rifles. "Carrier-tilt" is the term used to describe the offsetting motion of the bolt carrier group when rapidly struck by the operating rod. The problem that exists here is the accelerated wear on the aluminum receiver near the buffer tube. Most rifles utilizing gas piston operation have steel rails to provide support for the carrier during recoil. Steel receivers can handle the wear when they have had proper lubrication, but aluminum doesn't fare as well and also tends to heat up far more quickly. This excessive wear has proven to be a problem with several gas piston AR type designs.

The HK 416 has proven to be an excellent overall design. Cleaning time is far shorter than for the M16, and the service life of parts located in or near the receiver is extended due to far less heat being dumped into this area during operation. The heat buildup that arises from the direct impingement gas system is due to the hot expanding gases flowing down the stainless steel gas tube and into the receiver area to the carrier key. This heat buildup is what causes most early parts failure of the M16 series.[4] As many of the parts are heat treated, overheating quickly destroys the hardness of these parts, causing early failure. All of this was eliminated with the HK 416 through the use of its short-stroke gas system. Due to the success of this 5.56mm design, HK has recently introduced a nearly identical model in 7.62mm NATO, the HK 417.

The HK 417 uses a plastic magazine that is available in either ten- or twenty-round capacity. The shorter ten-round magazine would seem better suited to snipers due to its smaller profile, allowing for easy elevation adjustments without interference from the magazine. Since the HK 417 uses an AR pattern receiver, magazines are inserted straight into the magazine well as in the FN Mk17. Unlike the SCAR, the safety is not ambidextrous but is similar in operation to the M16, as is the bolt catch. Unlike the Mk17, the stock is collapsible in the same fashion as other AR pattern rifles. This stock has multiple locking positions. This is to better fit shooters of different heights or for the use of body armor, which generally requires a shorter stock for proper fit. The stock also has room for spare battery storage for whatever optics maybe in use on the weapon at the time. Like the Mk17, the HK 417 uses a full-length Picatinny rail system with attachment points on all four sides of the forend. The barrel can also be changed at operator level in a matter of minutes. Three different barrel lengths are available, 12 inches, 16 inches, and 20 inches. The gas system is similar to that of the HK 416. On the HK 417, however, the regulator has two positions, one for normal fire and the other for suppressed fire, as with the FN Mk17.

The weight of the HK 417 is more than that of the SCAR, with weight varying between just over nine and a half pounds and just under eleven pounds empty, depending on the particular model. Overall length depends on barrel length and stock position. The overall length of the HK 417 varies, with the short barrel measuring around 31¾ inches with the stock collapsed to just under 43 inches for the 20" model with the stock fully extended. The overall length is only reduced by roughly 3 inches when the stock is collapsed, as with the M4 series.

While having already been issued to several police and military units, the HK 417 is still being further developed and modified by its designers.[5] As with all new designs, including the Mk17, it will be some years before it can be considered fully combat proven. Given the excellent reputation of its little brother the HK 416, the 417 should prove to be a top notch replacement for the HK G3 that served so well for the past fifty years. With its aluminum receiver, it is doubtful that this rifle can achieve the service life of the G3, though few rifles will ever be able to make that claim. The G3 makes for hard shoes to fill.

In 2008–09, a semi-auto version of the HK 417 was introduced, known as the MR762A1.[6] This rifle is fitted with a cold hammer forged barrel, as are most other HK weapons. It offers a free floating four sided rail system, as with its selective-fire counterpart. It also has the same offerings with regard to accessories. Pistol grips of different geometry, butt plates with either convex or concave profile and an ambidextrous AR type safety lever are all available if desired. In addition, a flip up front sight is offered, as are detachable aperture sights that can be mounted to the full-length upper rail if iron sights are preferred to optics or as a backup should the optics fail, though this is a very rare occurrence with today's optics.

The MR762A1 uses a standard barrel of 16½ inches, although other barrel lengths may be offered in the future. This choice actually seems a bit short for a 7.62mm NATO caliber rifle, but it is just long enough to allow for keeping decent ballistic potential. A barrel in the 18"–20" range allows for a lower muzzle signature and increases muzzle velocity by two to three hundred feet per second. However, this does come with a decrease in handling quality and an increase in overall weight and length.

While the retail price for the civilian version is high, many collectors and shooters will likely be more than willing to pay such prices for the chance to own a long range, hard hitting rifle with the HK name on the side. The presence of these two letters preceding a weapon's designation usually means one of the world's best weapons, and this has been recognized as a truth since the late 1950s.

While the new generation of HK designs has veered off the path of HK's traditional roller-locked system, these new designs have built a good reputation of their own in recent years. The fact that both designs bear the same manufacturer's name should instill confidence in anyone trying to decide whether not to buy one of these new 7.62mm HK rifles.

One advantage of the HK 417/MR762A1 when it comes to field stripping is its similarity to the M16 series. The actual procedure for tearing the HK 417 down for cleaning is, in fact, almost identical. There are only a few operations that differ for the HK, and the primary reason for any variation at all is due to the change in its gas operation system. These will be covered as well as the reason for the differences.

Once again, clearing the weapon is first priority. The magazine removal for the HK 417 is identical to that of the M16 series. The 417 uses a push button type release, which is located on the right side just above the forward portion of the trigger guard, as with the M16.

Once the magazine is removed, clear the chamber by pulling back the charging handle, which is located directly at the rear of the receiver. The 417 has locking latches on either side of the charging handle, again much like that of the M16 (locking latch on one side only). This lock is to ensure that the charging handle remains in its forward position when not in use. Visually ensure that the chamber and magazine well are clear before proceeding.

If these areas are clear, safety the weapon. The safety is actually one feature where the HK differs from the M16 family. The HK safety can and in fact should be applied without first cocking the weapon. On the M16 series, the hammer must first be cocked before the safety can be applied. This is actually a nice added feature, but a weapon should never be considered safe until it is manually cleared and visually inspected. Allow the charging handle to slowly ride forward after you inspect the chamber area. Disassembly of the HK requires the use of a unique tool, which is housed within the buttstock. Access to the storage compartment is gained by rotating the recoil pad of the buttstock ninety degrees clockwise. This tool is needed to remove the special recessed takedown pins of the 417. With the M16 series, these pins can be removed with the point of a bullet or any other device of appropriate size.

The takedown pins on the HK are captive, as with the M16, and they're not meant to be completely removed. The takedown tool is used to first push the rear takedown pin to the right. This process is repeated for the front pivot pin as well. After these pins are moved fully to the right, the upper and lower receivers can then be separated.

At this point, the charging handle can be withdrawn without spring resistance. This is because the action's recoil spring is housed within the buffer tube located at the rear of the lower receiver. When pulled back far enough, the bolt and carrier group can be grasped and pulled free from the charging handle. The handle can then be pulled out and up until it is free of its engagement with the upper receiver. A captive firing-pin retainer is drifted out to the left of the carrier group. Before the firing-pin spring can be removed, a unique safety must first be lifted. This is located within the recess of the bolt carrier. The firing pin and spring are then withdrawn from the rear of the carrier. The bolt cam pin can be pulled up and clear of the carrier and the bolt can then be pulled free from the front of the carrier.

The rail retainer screws are located at the lower right portion of the rail system. The MR762A1 manual indicates that a 5mm hex key is required. Loosen and pull these screws out as far as they will travel. These are also captive screws and are not meant to be completely removed. Once screws are pulled to their travel limit, the rail can be pulled forward and free of the receiver.

The gas piston can be taken apart by pulling back the piston rod until it is free of the piston. The piston will need to be held in place to do this. Once clear, the piston rod and spring can be shifted slightly to one side and pulled free. The piston can then be pulled to the rear and free of the gas block.

The final step is to depress the buffer/spring retainer at the front of its tube located in the lower receiver. The buffer and spring can then be released slowly and pulled out from the buffer tube.

Reassembly is simply the reverse of the takedown procedure, though some steps are crucial. Proper bolt alignment is necessary to reinstall the cam pin. The firing-pin safety will have to again be lifted to reinsert the firing pin and spring. Once the pin and spring have been reinstalled, push the bolt in and lift the safety lever while holding the carrier vertically with the bolt side facing up. The firing pin and spring should not fall out.[7]

Those familiar with the M16 takedown procedures will have recognized a great number of the steps involved in tearing down the HK 417/MR762A1. The same will apply to the maintenance regimen for both weapon systems.

Those schooled in caring for the AR pattern rifles will do just fine with the HK 417/MR762A1 series. As many of the components used in both designs are nearly identical, at least in design if not in dimension, many of the cleaning tips and techniques used in maintaining the M16 will apply to the HK 417 series.

The most crucial areas to attend to on the M16 design are the bolt/carrier group, chamber, and barrel extension areas, along with the gas tube. Of course the gas tube of the M16 series has been replaced by a gas piston system on the HK. HK does recommend a light coat of oil for the gas piston and rod. The suggestion to lube this particular area varies from one firearm manufacturer to the next. If the maker suggests a coat of oil or grease be applied, then there might well be a valid reason for this. Oftentimes it is just to help reduce parts wear. The actual functional use of lubricant in this area is questionable, as the heat buildup in this area is rapid and extreme, with temperatures sometimes reaching over 300 degrees Fahrenheit. Any oil will likely be burned off quickly. High-temp bearing grease may do somewhat better in this regard, but the high viscosity of this choice of lubricant may adversely affect the reliability of the system. If the design calls for a dry gas system, then it is best to follow the manufacturer's recommendation, as with the M14.

As a general rule, a light coat of oil is best on most components unless the specific area calls for a heavier level of lubrication. As with the M16 series, the trigger group should receive a light coat of lube only on areas where metal friction wear is obvious. In sandy conditions, a minimal application of oil will help keep grit out of this crucial area of the weapon. This is important to remember, as this is one of the locations most prone to contamination from sand. The large number of closely fitted parts here can easily be locked up from a small amount of sand or grit mixing with oil. Keeping this area as dry as practical will help prevent the trigger group from locking up or sticking. This rule of thumb should be applied to all firearms intended for use in desert conditions.

Aside from the trigger group, other areas of the lower receiver that should receive a light coat of lube are the buffer and spring assembly and their retaining catch, mainly the buffer and spring interface. Oil in this area helps with heat buildup due to friction. Again, in sandy conditions less is more. This lube should be concentrated around the buffer itself and only at obvious wear points. Friction wear points will be clearly visible after only a relatively small number of rounds have been fired through a weapon. These wear points are an excellent guide for the proper application of lubricant to help minimize wear without going overboard on the oil.

The maintenance procedures for the upper receiver area are far more involved, however. Here areas of concern are many. The bore and chamber are always crucial components for the AR style bolt system and this is one area that should not be neglected. This is one of the key areas of this design that must be kept clean if reliability is desired. As the HK uses a similar style bolt lock-up here, this should, by default, apply to it as well. The area of particular concern here is the locking lug recess of the chamber, along with the corresponding lugs on the bolt. The large number of small lugs and lug recesses makes cleaning this area time consuming, but proper cleaning of this area is an absolute necessity. This cannot be overstressed. For the M16, special patches are issued for this purpose. These patches are cut in the same geometry as the lug layout, with little teeth around the circumference. In actuality, any patch can be used along with a small non-marring pick to help remove grit. Toothpicks are great here. This area requires a great deal of attention for several reasons. First, the large number of small lugs, while facilitating bolt locking and unlocking with minimal rotation, also rely more heavily on that small degree of rotation for proper functioning. Any excess of buildup here is more likely to impede proper bolt operation. Second, this area tends to get very dirty due to the large amount of fouling generated by the direct impingement gas system as used on the M16. With the HK, this area will not get nearly as dirty, but the large number of lugs is still present and focus in this area is still needed.

The feed ramp area is also a key area for maintenance attention. This area must be kept free of buildup from both carbon and metal deposits that result from bullet jacket material rubbing off due to cartridge feeding. An excess of this buildup can increase friction of the cartridge as it is being guided into the chamber. This increased friction will slow the carrier momentum and could prevent proper bolt lockup. While both designs utilize a forward bolt assist to help ensure bolt locking before firing, using the forward assist after every shot is not practical or even a possibility, as this would more than likely result in a great many dead troops, as they would basically be using a weapon akin to a manual repeater. The chamber/bolt area of the HK 417 will not get anywhere near as dirty as that of the M16. This is of course due to the change in the type of gas system, from a direct impingement to a short-stroke gas piston system. Also, due to this change in operation the remaining areas of the upper receiver will remain far cleaner on the 417. The carrier of the 417, which is considerably different in purpose, if not necessarily in layout, will not receive the volume of expanding gas and its consequent carbon deposits. As a result, the cleaning time needed in this area will no doubt be considerably shortened. The bolt itself will likely not receive the amount of carbon buildup that stems from a direct impingement gas system. This does not change the earlier requirement for chamber and bolt cleaning, however, as this area still tends to receive some carbon depositing.

The charging handle, which remains stationary during firing on both designs, should receive a light coat of lube, primarily at the front hook where it engages the bolt carrier. Any other areas of this part showing signs of wear should also be lightly lubed.

The gas system on the HK will require a considerable amount of attention during cleaning, as it is this area that will receive the brunt of the carbon buildup and fouling during operation. The gas piston and gas block, in particular, are the locations most often associated with carbon buildup due to prolonged weapon firing. For cleaning this area, pipe cleaners, again, have long been perfect for cleaning the gas port and small diameters associated with many gas systems. A special, extra-long pipe cleaner has been available for years and was specifically made for cleaning the gas tube of the direct impingement M16.

Other tools most often used in cleaning this area include stiff nylon brushes, and for severe fouling issues a bronze or brass brush will provide extra scrubbing action. These metal brushes are best used sparingly, as this will help maintain a better finish on parts. As always, these metal brushes must never be used on the aluminum components utilized so heavily in these designs.

If these basic maintenance techniques are followed, the HK 417/MR762A1 series should function reliably under any and all conditions. That is of course within reason, as few designs have proven 100 percent reliable in very sandy conditions, with the AK based designs being the most usual exceptions in this regard.

The 417 has a close relative in the semi-auto only G28, which shares a majority of the same components. This rifle has recently entered service with the German army as its new DMR. The primary difference between these two rifles, other than the lack of a full-auto option on the selector, is the steel upper receiver of the G28. Steel was no doubt used to enhance the accuracy of the weapon, as it offers greater rigidity and also acts as a better heat sink than aluminum for sustained fire. The G28 is being used in Afghanistan by German troops.[8]

9

Galil ACE

The final rifle to be covered is one of the newest and has only been introduced in the last few years. Nonetheless, it has already been adopted by its parent nation, Colombia, as well as a nearby neighbor to the north, Guatemala.[1]

The Galil ACE is manufactured by Indumil, the state arsenal of Colombia.[2] Israeli Weapon Industries (IWI), Ltd., no doubt helped in this regard, as they are the original designers of the Galil rifle series.

The Galil ACE is offered in three calibers: 5.56mm NATO, 7.62mm NATO, the 7.62×39mm Soviet. These models differ primarily in barrel length. In addition, there's a special 5.56mm short version designated the "N" model. This version differs from the other 5.56mm models in that it uses an M16 pattern NATO standard magazine while the other 5.56mm variations use the original Galil pattern magazines.

The 7.62×39mm Soviet models 31 and 32 wisely use the standard AK-47 magazine pattern, which is likely the world's most common rifle magazine, as there are more AKs around than anything else.

For our needs the models 52, 52L, and 53 are of particular interest, as these are the "battle rifles" of the ACE line. The 7.62mm NATO Galil ACE models use the standard 7.62mm NATO Galil twenty-five-round magazine. Let's begin with an overview of the Galil ACE rifle in general and discuss its improvements and changes over the original Galil series rifles.

The Galil ACE has brought with it a great many changes but has kept the heart of the Galil rifle: its AK based long-stroke gas-piston-operated action. To begin with, a Picatinny top rail is standard, as are side and bottom rails on the forend. Grip panels are normally installed on the side and bottom rails unless accessories are needed. These grip panels provide a more comfortable hold on the forend as well as decreasing the chances of hand slippage.

The iron sights are similar to those of the original Galil, but the front sight post is now protected by a set of wings rather than the circular hood used on the original Galil rifle. The rear sight is still of the original aperture pattern and is still located at the rear of the receiver cover, giving these rifles a much improved sight radius over standard AK pattern rifles, which have the rear sight mounted on the gas block in front of the receiver. This generally means improved accuracy for the average user, in addition to much faster target acquisition due to the use of an aperture rear sight.

Perhaps the most noticeable changes are in the stock and lower receiver area. The standard ACE stock is a telescoping model very much like that of the U.S. M4 carbines series. An optional folding version is reportedly available. This is likely similar to that of the original Galil rifle. The biggest change in the receiver is that it is no longer a one piece unit but now consists of upper and lower sections and the bottom section is made of plastic for a big savings in weight.

The safety/selector lever is still ambidextrous, but gone is the large AK type safety lever/dustcover on the right side of the receiver. In its place is a much smaller thumb lever, with its counterpart present on the left side of the receiver above the pistol grip, where a safety is best located for right-handed users. Once again the left side safety moves to the rear to allow fire. This is similar to the original Galil left side safety/selector.

The right side, vertically oriented cocking handle of the Galil is gone and has been replaced by a left side mounted, horizontal cocking handle that would be very familiar to FAL users. With the ACE, however, the cocking handle reciprocates during firing, as with the FN Mk17. This is because the cocking handle is connected directly to the bolt carrier. To prevent dust from entering the resulting cocking handle slot on the left side of the receiver, a spring loaded dustcover is installed which returns to position after every shot. This arrangement does allow the cocking handle to be used as a forward assist if needed to guarantee bolt lockup in the event of an excess buildup of fouling. This arrangement is not quite as convenient for left-hand users as was the cocking handle of the original Galil.

Some may wonder why a telescoping stock was installed in place of a more compact folding version. The answer is recoil control. The new stock is supposedly better at dealing with the recoil forces generated during firing.[3] As the ACE is a military weapon, it is selective fire and the cyclic rate remains unchanged from that of the original Galil, at roughly 650 rpm.

As mentioned, the Galil ACE has been adopted by the military forces of Colombia as well as those of Guatemala, with more countries likely to follow suit in Latin America.[4]

If the Galil ACE proves as reliable as the original Galil, it should prove to be one of the best battle rifles available, though the Chilean 542–1 and the new SIG SG751 are also rifles to beat. It must be remembered that the FAL and G3 are still in production in parts of the world and are still contenders in this regard as well. The military FAL is still being cranked out in Brazil and the G3 is still coming out of Turkey and Pakistan. However, the Galil ACE is far lighter than the G3 and somewhat lighter than the FAL, though the FAL 50.63 carbine is a fairly light 7.62mm in its own right.

As discussed, there are three variations of the 7.62mm NATO Galil ACE rifle. The first is the model 52, which uses a 15¾" barrel similar in length to that of the original 7.62mm Galil SAR. The short variation weighs in at just over eight pounds empty and measures a bit over 33½ inches with its stock collapsed, as with the M4 series. Add 3 inches to this for the extended length. The slightly longer model 52L uses an 18" barrel, which adds a few ounces to the weight. Overall lengths are increased by 2 inches respectively. The long ACE version is the model 53, which uses a full-length 20" barrel and weighs just under eight and one-half pounds empty.[5]

It is interesting to note that Israel has chosen not to manufacture this new rifle locally. This actually shouldn't be that surprising, since for the last years of its production the original 7.62mm Galil was manufactured in Colombia as well, under license from IWI, of course.[6] IWI no longer offers the original Galil in 7.62mm, but the sniper model and several 5.56mm variations are still listed in their product catalog. For the last decade, IWI has instead focused its attention on the 5.56mm Tavor series bullpup, in its many variations. While the Tavor had some initial minor issues, it is easily considered more reliable than the M16 rifles series and has seen fairly heavy use in Israel. The Israelis no doubt have had to adjust their tactics to compensate for the shortcomings of the bullpup configuration, though since most combat conditions faced by Israeli troops have been urban in recent years their decision to stick with the 5.56mm round makes perfect sense. Israel does have several types of longer-range 7.62mm NATO weapons available when conditions call for their use. The 7.62mm Galil sniper rifle, stockpiled 7.62mm standard Galil rifles, MAG machine guns and new Negev NG7 belt-fed machine guns are all available to Israeli troops if needed.[7]

The new Galil ACE, however, offers a much lighter weight than the original, in addition to coming standard with many of the current popular features that the original Galils did not possess without either heavy modification or a considerable increase in weight when retrofitted with such accessories.

One feature present on the new Galil ACE that was missing in the original Galil is of particular benefit. The new Galil ACE is equipped with a last round bolt catch, as are many modern rifles. While many may consider this feature unnecessary, not having to guess whether or not your weapon is empty or has misfired can save precious seconds in a time of crisis. Training can of course eliminate the natural tendency for one to stop and attempt to determine what the problem is, but the distinct feel and sound of a last round bolt catch activating eliminates the need for such training. The fact that most new rifle designs include a bolt hold open feature is evidence that more military forces are learning to appreciate the tactical advantages offered by this device, especially when combined with a quick release lever or button to drastically cut down on reloading time. With a bold hold open feature the decreased time in reloading can mean the difference between life and death. Again, the advantage gained here is small and can be overcome with proper training. However, any additional training takes away from the military budget and the U.S. isn't the only nation feeling the economic pain that has existed for several years now.

While the Galil ACE is still quite new and must be considered as yet relatively unproven in combat, its adoption by both Colombia and Guatemala will ensure that it will see operational use in short order given the general civil instability that exists in both nations. Given the widespread popularity of the original Galil in Latin America, we have every reason to expect the Galil ACE to see similar popularity to that enjoyed by the original. This is especially true due to the fact that Indumil had the forethought to keep the guts of the Galil when designing the new rifle. The end result is a highly developed AK-47 with a lot of extra features. It just happens to be an AK type rifle that packs a whole lot of power due to its 7.62mm NATO chambering.

The Galil ACE is such a new design that I was unable to obtain a copy of its operator's manual, though takedown procedures appear to be quite similar.

Once the weapon is cleared, the recoil spring guide lug located at the rear of the receiver cover is depressed to allow removal of the receiver cover. The cover is then pulled slightly up and to the rear until clear of the forward retaining lip. Next the recoil spring guide is pushed forward until its lug is clear of its guide. This can then be pulled slightly to one side to allow it to be withdrawn to the rear and removed.

Aside from the rail attachments sitting atop the gas tube, there appears to be little difference between this component and that of the original Galil. Aside from removal of the gas tube, there should be no further takedown of the weapon.

Maintenance procedures should follow those of the original Galil quite closely. As for the new synthetic lower receiver of the Galil ACE, it appears that this piece is not meant to be separated from the steel upper for any reason. There is not much available information on the new ACE yet regarding maintenance specifics on this new receiver design, so it is unclear if any damage will occur should some especially harsh solvents be used for cleaning purposes. If standard weapon cleaning solvents are utilized, there should be no problems.

10

Post–World War II Combat Cartridge Development and Performance

While the preface of this book made clear the reason for the recent renewed interest in the long range power of the 7.62mm NATO cartridge, we have not really covered the actual differences between the various cartridges in common use today. This chapter is going to be rather lengthy and will be constructed in chronological fashion, primarily covering the metallic cartridge era up to the present day, with a focus on the newest developments and why we are seeing these developments. There will also be some basic explanation regarding ballistic data and what it means in its use when comparing cartridge performance.

While we will primarily be covering the development of metallic cartridges, a brief history of cartridge development in general is in order and is best explained from the start. Firearms development really begins with the introduction of gunpowder in Europe after the transfer of technology from China in the thirteenth century. This is just a rough estimate, of course, and to this day there remains a great deal of debate as to how exactly this technology made its way to Europe and where it first occurred. While the first European mention of gunpowder and the various weapons that used it didn't show up in the historical record until the early to mid–1200s, its use in Asia dates back much further. By the time gunpowder first made its appearance in Europe, the Chinese had already been using it in battle for over three hundred years. After the Mongol invasions of China, these nomadic horse warriors brought this weapon with them and used it in their other endeavors.[1]

After seeing the effects and usefulness of gunpowder and firearms in their battles against the Mongols, Europeans wasted no time learning to harness this new technology. While Western long guns made several short strides of advancement after this first introduction, none were major breakthroughs. From the thirteenth century until the early nineteenth century, long guns never really advanced beyond being single-shot weapons that were slow to reload. The various changeovers from matchlock technology to wheel lock technology and finally to flintlock never brought any great tactical breakthroughs. This is of course aside from the elimination of the glowing red embers from the match lock's fuse (match).

However, by 1840 the percussion cap was widely introduced for use in igniting the main powder charge of single-shot weapons.[2] This is where the story turns a corner and the birth of modern firearms truly begins. In 1800 Edward C. Howard discovered a fulminate compound that led to the development of the percussion cap shortly afterward. In 1807 a Scottish preacher, the Reverend Alexander J. Forsyth, developed a concussive material that was to be the basis for percussion priming. The compound he developed was comprised of fulminate of mercury, chlorate of potash, charcoal and sulphur. This concoction was easily ignited when crushed with a sharp blow, and its development was the true birth of the modern cartridges used today.[3]

While who deserves credit for the percussion cap itself is often disputed, the Reverend Forsythe deserves the greatest credit, as he developed the compound that made the percussion cap possible several years later. Despite the various claimants to the percussion cap's invention, it was Forsythe's love of hunting that encouraged him to develop a stealthy means of igniting the powder in his shotgun. Prior to this, the shot of a firearm was always preceded by the spark of the flint, which had the disturbing effect of alerting the game animal.

In any case, it took some years for the military establishment to embrace the new technology of percussion priming. This is really nothing new for a profession that has a long history of always fighting the last war and being skeptical of new technology. At any rate, by 1834 the military establishment had generally accepted the percussion system as reliable technology. Following this in 1835, the fragile pinfire system was patented. This was a very close relative to the modern self-contained metallic cartridge.[4]

The first modern metallic cartridge was the BB cap, which was nothing more than a newly developed copper percussion cap with a lead ball inserted in the front. This was the true birth of the rimfire. The .22 rimfire has historically been the most used cartridge in the world.[5] While rimfire rounds were the first successful metal cartridges, much more powerful centerfire cartridges would soon evolve. While centerfire rounds first appeared in the mid–1860s, within ten years many different variations had been developed.[6] The venerable .45-70 Government cartridge became standard in 1873.[7] That same year saw the introduction of the "gun that won the West," the 1873 Winchester lever action rifle. This rifle was made in several calibers, but by far the most well known was the .44-40 Winchester Center Fire (WCF). Both the gun and the cartridge it used became very popular for use on the frontier, as the .44-40 was also capable of being fired from the Colt "Peacemaker," also known as the model 1873 Single Action Army revolver. This revolver was adopted by the U.S. Army in 1873 in caliber .45 Long Colt (LC).[8]

While the U.S. Army and the Western cowboys had their favorites, the buffalo hunters had their own long range favorites as well. The legendary Sharps model 1874 was one of these long range rifles and was chambered in a variety of powerful black powder cartridges with names like the .50-90 and the .44-77.[9] These were designations for some of the many black powder cartridges, and while most are familiar with the terms, it goes like this: the first number indicates the bullet diameter and the number following the hyphen indicates the number of grains of black powder loaded into the cartridge case.

While these rounds may be considered obsolete today, it must be remembered that these were the top dogs of the day. In fact, the .50-90 is often considered the primary reason for the American bison nearly going extinct,[10] that and the desire of a small, select group of Americans to cut off the primary food source of the Plains Indians, as they were viewed by this small group as a major obstacle to American expansionist views.

By the late nineteenth century, ammunition development had come close to today's current level of technology, but there was still room for improvement. Black powder was effective but left a telltale cloud upon detonation. This will give away the shooter's position and obscure his vision, affecting any follow-up shot. An improvement had to be developed. Smokeless powder first appeared near the close of the 1800s.[11] With its advent came a new series of popular cartridges, and some earlier popular black powder cartridges fell by the wayside. These were usually the ones that were unable to cope with the higher pressures developed by the new type of powder. Some cartridges did successfully make the transition and are still relatively popular today. The .45-70 Government is a prime example of this.

Along with the new type of powder and the new cartridge designs came the new pointed spitzer type bullet. This streamlined projectile brought with it a far superior aerodynamic geometry along with exterior ballistics far beyond anything previously seen.

One problem with the new powder was that it required greater ignition properties from the old mercury based primers. These old primers also tended to be unstable and deteriorated over time, which sometimes led to misfires. In addition, they were highly corrosive and would rust the firearm in short order if it was not thoroughly cleaned after firing. Newly developed noncorrosive primers were in use prior to the Great Depression era, though military production continued using corrosive primers, which are still encountered even today in some surplus ammunition.[12] This is especially true of surplus ammunition that originates from the former communist bloc nations.

With these newer ammunition developments came yet another new series of popular cartridges, many of which are still popular today, and their names are often immediately recognizable. Calibers like the .30-06 (the '06 here being its year of introduction) Springfield and the .357 Magnum (introduced in 1935) became the new top dogs of their day. The big .50 Browning Machine Gun (BMG) cartridge had been developed following the end of World War I. Many of the calibers from this era are quite popular today, and for years military adoption of a particular cartridge all but ensured commercial success as well. Consequently, whenever there was a search for a new weapon system or caliber competition was naturally very tough.

Black powder was man's only known explosive for most of our history. The more modern high explosives didn't appear until well into the nineteenth century. While it allowed firearms to develop to a certain point, it was primarily the industrial revolution that opened the door to design advances in weapons technology. Prior to the 1860s, single-shot muzzle loaders were the norm. There had been earlier attempts at breech loading, but none were very meaningful. The problem was in the lack of any method of sealing the breech to prevent gas leakage during firing. The advent of the copper (later brass) metallic cartridge provided several solutions at once.

The advent of the metal cartridge case gave shooters the means to carry a convenient, self-contained round that was all but weatherproof. It was easy to carry and quite rugged with regard to transportation and shipping. This made it safer for logistics purposes as well as it was far safer to ship than black powder in kegs. The centerfire primer was to make cartridges even safer than the original rimfire designs. Most important, though, in ammunition development terms, the metallic case itself was an effective means of sealing the breech during the firing sequence. The pressures created by the expanding gases within the cartridge swelled the case before much gas could leak out.

With the major long-standing issues concerning breech loading solved, many new roads were now open to small arms design.

Repeating weapons were not new by this time, but prior to the development of the metallic cartridge this concept usually took the form of some sort of revolver.

The idea of magazine repeating weapons soon followed. The Volcanic pistol was one of the first. This design introduced the under barrel spring loaded tubular magazine, which was the birth of the Winchester legacy.

Once this new form of ammunition became commonplace, it didn't take long for the really quick men to come up with some new ideas. None was quicker than John Moses Browning. He proved his genius on more occasions than any other small arms designer before or since.

Although Browning did develop some single-shot weapons, he is best remembered for his vast number of repeating designs. These began with the Winchester 1886 lever action rifle, which introduced an improved bolt locking system that offered far greater strength than its earlier brother, the model 1876, a design based on the original Henry lever action and subsequent model 1873, the "gun that won the West."

This was not Browning's first design for Winchester, merely his first of many repeaters. This relationship with Winchester continued for many years. It ended with a falling-out over the Auto-5 shotgun, the first and, for many years, only truly reliable self-loading shotgun.

Browning's work for Winchester survived the industry's transition from black powder to smokeless. In fact, the first Winchester lever gun design to use smokeless powder was another Browning model, the famous 1894. Many consider this the finest traditional lever action ever made. The original cartridge used by this model was the .30 Winchester Center Fire (WCF). This round is perhaps better known as the .30-30. The black powder designation was instituted more to allow other makers to market the same caliber without legal issues. While it is not considered a real performer by today's standards, the .30-30 was a hot number in its day. As smokeless powder creates much higher pressure levels than black powder, careful work had to go into the design of a successful cartridge case. The .30-30 proved quite successful to be sure. It is the all-time most popular lever action cartridge.

While the industrial revolution may have helped trigger the rapid advances in small arms design, it also illuminated the shortcomings of propellant technology at the same time. The chemical formula for black powder had been known since the ninth century, and while it proved an effective propellant, it was not without its shortcomings. Black powder is not the ideal propellant for repeating small arms mechanisms. Black powder gives off a tremendous amount of smoke when it burns, but this is a tactical shortcoming. This characteristic was also one of the motivating factors when a search began for a better alternative propellant.

The issue regarding black powder and repeating firearms has to do with the burning characteristics of the powder. Black powder is not the cleanest burning propellant, and the residue it leaves behind has a tendency to attract moisture.

What the industrial revolution did for small arms technology was open up an endless number of roads for new design. Parts could now be made with relative precision and consistency in large numbers rather than by hand as before. And this allowed designers to experiment with endless varieties of feed systems and firing mechanisms. It was actually a combination of the industrial revolution with the introduction of the metallic cartridge that allowed for this explosion in repeating small arms design.

With the ability to produce consistent parts that are fairly quick paced, large manufacturing facilities began to appear that could rapidly produce these new designs. While Colt had been making revolvers since 1836, these did not begin to utilize metallic cartridges until the 1860s. The first repeating long guns began to show up at roughly the same time; again the metallic cartridges production was the final part of the equation.

The first successful models were the .56 Spencer carbine and, more important, the 1860 Henry .44 rimfire lever action rifle. This latter design would lead to the legendary Winchester lever actions. The Winchesters were *the* repeating rifles of their day. The Henry, in fact, proved to be more than durable enough for military service, but never was procured by the military in significant numbers. The Henry was often purchased by individual Union troops and small units. It proved its worth with a reliable mechanism and a sixteen-round under barrel tube magazine. This gave the Henry a huge advantage over the muzzle loading rifled muskets that were the most common weapons of the Civil War. The repeating rifles used during the war quickly gained a good reputation despite their limited use in combat. By the war's end, it was clear that repeaters were here to stay and that the single shot's and especially the muzzleloader's days were numbered.

With the Henry rifle proven in combat, it didn't take long for the design to see some improvements. The Henry rifle was effective but not without its faults. It had no forearm to protect the shooter's off hand from a hot barrel. Also, its method of loading by first compressing the magazine follower, then twisting the barrel, then loading rounds in from the muzzle, left a great deal to be desired.

The Winchester 1866 was born out of the combat experience with the 1860 Henry. This new Winchester used the same basic lever action but added a wooden forearm to help shield the user from heat coming off the barrel. More important, a new right side loading gate (King's loading gate) made the rifle far easier and quicker to load. The cartridges could now be simply pushed forward and down into a covered spring loaded gate (located at the receiver end) and into the tube magazine, which was still located beneath the barrel. This new system was not only faster and easier to use when loading the weapon but also far more reliable. The original Henry had an open magazine tube. What this means is that the loading system used a slot cut the length of the tube. The magazine follower had to be compressed by hand, which required a slot running the entire length of the magazine. This slot was an easy access point for mud and grit even if it was kept completely dry and free from oil. One drop in wet soil and the magazine would be packed with mud. When shooting starts, most men tend to go prone out of instinct alone. This is when the problem really begins. It is nearly impossible to keep a weapon free from dirt or sand when operating in the field.

The 1866 Winchester eliminated this issue entirely by using its side loading gate and an enclosed magazine tube. The model 1866 still used the original .44 rimfire cartridge introduced with the Henry rifle. This cartridge was a bit underpowered even for its day. It still did the job, but there was considerable room for improvement. This cartridge produced a bullet moving at roughly 1,100 feet per second. By comparison the Spencer carbine used a 350 grain bullet moving at nearly 1,200 feet per second. This was the Henry's closest competitor at the time. The Spencer carbine held only seven rounds in its tubular magazine. The Spencer's magazine, however, was located within its buttstock and was better protected than that of the Henry. The 1866 Winchester was available in either a long barrel rifle variation or a handier carbine model. While the link action could handle the .44 rimfire Henry cartridge, there was a new series of centerfire cartridges beginning to hit the market. This toggle link action proved strong enough to handle these as well, up to a point.

The next big improvement in repeating rifles came with the famous 1873 Winchester. The new model 1873 introduced the now famous .44-40 centerfire cartridge. This new cartridge used a tapered case design that suffered fewer feeding issues than straight walled cases like that of the .45 Long Colt (LC). The .45 would have made the perfect companion weapon to the U.S. Army's new Colt .45 revolver that was introduced at roughly the same time. A rifle utilizing the same cartridge as a handgun was a great logistical help, but the .45 would not successfully be chambered for a lever action rifle until years later, well after the frontier had closed. Due to the mechanical issues of chambering the .45 in a rifle, the .44-40 was to become this combination cartridge of choice. The Colt Peacemaker was the Army's revolver at the time, and due to its rugged design and reliable operation it soon became the premier handgun of its day, though Smith and Wesson and Remington might disagree here. As the Peacemaker was such a popular handgun it was soon chambered in the .44-40 to make the perfect companion sidearm to the 1873 Winchester. This model was introduced in 1878 and for a time Colt even went so far is to mark these peacemakers with the words "Frontier Six Shooter." These early Peacemakers are currently some of the most valuable revolvers to collectors.

Though these weapons may seem quaint by today's standards, these were the AK-47s of the era. The Sioux, Cheyenne and Arapaho clearly demonstrated the capabilities of the Winchester during the Battle of the Little Bighorn in 1876. If there was a weakness in the design of the 1873 Winchester, it was in the power of the cartridges in which it was chambered. The most powerful chambering for this model was the .44-40, which was at best a 200-yard round. An expert might be able to extend this range a bit, however. The limit of the toggle-link action was the reason for the relatively weak chambering offered in the 1873.

The larger model 1876 was introduced and this rifle was beefed up to help it handle the more powerful cartridges it offered. Cartridges for the new 1876 Winchester included relatively large black powder rounds like the .45-60, .45-75, and .50-95 "Express" cartridge. While these were far more powerful than the chamberings offered in the 1873, they weren't quite up to the same level of performance as the .50-90 Sharps buffalo guns. It was these powerful buffalo rifles with which Winchester was hoping to compete in its introduction of the 1876 model. It was clear they would have to have a stronger design.

They designed such a rifle in 1886 by enlisting the help of John Browning. The new model 1886 could finally handle the popular and powerful .45-70 Government cartridge. It could also handle the larger buffalo rounds. The .50-110 Express was one of the more powerful chamberings offered in the model 1886, and this model was finally selling to the buffalo hunters just as the buffalo began to vanish from the American West, although it is unlikely that many would feel too bad for Winchester in this regard.

The new rifle utilized two vertical locking bars attached to the lever to increase strength during firing. These bars helped secure the bolt during peak firing pressures. It was this addition that allowed for the chambering of much more powerful cartridges.

The next Winchester model developed was basically a downsized version of the 1886. The model 1892 was to become a quite successful design and sold well into the twentieth century, making the transition to smokeless powder cartridges.

The final lever action tube fed Winchester design was probably the most famous, the Browning designed Winchester model of 1894. This is the biggest selling lever action of all time by far, with production numbers of over 7 million. While the 1894 was offered in black powder cartridges, it was one of the first production rifles to use stronger steels, which were introduced to handle smokeless powder pressures. The 1894 made this transition early on and the .30 Winchester Center Fire (WCF) was one of the first smokeless powder cartridges to become a big success in this classic rifle. This cartridge is of course better known as the .30-30 and was one of the most common chamberings for the 1894 model.

The model 1894 would become the premier deer/black bear rifle for much of the more heavily wooded regions of North America. Wherever shooting distances were limited to 100 to 150 yards, the .30-30 model 1894 would no doubt be found in numbers. The Savage 99 lever action was one of its few serious competitors, along with competing models from Marlin, another big lever action name in American rifles. Neither of these company's models would ever see production numbers anywhere near those of the 1894, however.

The one major flaw with these tube fed lever actions, if it can be considered a flaw, was the inability to safely use the new spitzer bullet design that was beginning to become the most popular military projectile pattern due to its much better drag characteristics. These fast handling Winchester and other traditional lever actions could not use these new style bullets because the under barrel tube magazines stored the ammunition in line. Storing the rounds in line meant that a pointed bullet could potentially detonate the primer of the cartridge directly in front.

It was for this reason that Browning once again came to Winchester's aid. He designed a lever action rifle that stacked cartridges vertically rather than storing them end to end. In doing so, Browning was able to use the newest military cartridges that were more powerful and had far better long range ballistics due to their use of the spitzer bullet design. These bullets were a great deal better with regard to bullet drop at ranges over 200 yards and this would allow the use of lever actions for hunting in the West where large pronghorns and mule deer were quite numerous at the time.

While the 1895 Winchester that Browning designed would not be considered to have the strongest action of all time, its rear locking bolt was more than strong enough to handle the more powerful smokeless powder military cartridges like the .30-06 Springfield. The model 1895 was not the only new lever action to take the approach of storing ammo in a safer arrangement. The Savage 1895 and later 1899 used a similar layout. The Savage introduced a rotary magazine and this design proved to be the primary competition for the 1895 Winchester. While the majority of the world's armies were switching to new bolt action designs, some military purchases of model 1895s did take place. Russia made a purchase of the 1895 for the imperial army in its own caliber, the 7.62×54mmR. Though the lever action had proven the worth of the repeating rifle, it was not to become the dominant military action. Ottoman Turks had proven the value of the Winchester during their war with Russia in the late 1870s. Here the Winchester easily outgunned the single-shot Russian rifles, which was the stimulus for Russia's quest for a repeating rifle.

The first bolt action designs appeared many years before the introduction of the Winchester lever actions. One of the first was the von Dreyse, which was introduced in the 1830s. These early designs would also have to wait for the perfection of the metallic cartridge to achieve their true potential. The later bolt action designs that began to show up were to become the military's weapon of choice.

Probably no weapon represents the bolt action concept better than the 1898 Mauser (M98). Many consider the Mauser 98 to be the best overall bolt action design ever. The 98 was not the first Mauser bolt design but was the final "perfected" variation in a series of bolt action rifles developed by Peter and Paul Mauser over a number of years.

There were other good bolt action designs, like the British Enfield and the Krag-Jorgensen from Norway. The Mauser, however, seemed to offer everything. The U.S. Springfield M1903 was heavily based on the 98 Mauser design and the U.S. paid for the permissions. The Japanese Arisaka was also based on the 98 action.

The bolt action was to become the dominant military weapon for much of the first half of the twentieth century. They are still used a great deal today for sniper rifles due to their inherently strong, rigid lockup, which tends to translate to better accuracy, which is crucial for an effective sniper weapon. They are much slower to fire than a semi-auto, but overall accuracy is usually better. There are exceptions to this of course, as there have been some remarkably accurate semi-auto sniper rifle variations on battle rifles like the M14 and G3, specifically the M21 (sniper variant of M14), and the G3/SG-1 (sniper variant of G3A3).

While the United States had adopted the Krag in 1892, this well-made and -finished side loading design was not to see the same level of success as some of the other designs that were coming out at this time. In fact, the model 98 Mauser would eventually become the second most heavily mass produced firearm of all time, with numbers of nearly 50 million. Only the AK-47 would ever be produced in larger numbers.

The British Short Magazine Lee-Enfield (SMLE) would also end up seeing a considerable level of success, though not produced in numbers approaching anything near those of the Mauser. While many consider the SMLE inferior to the Mauser, it did prove to have one big advantage over the model 98 in that it used a ten-round magazine, which was to give British troops a huge firepower advantage over the five-shot Mauser equipped German troops. While practiced troops became quite adept at reloading their rifles while on the move, it still requires a break in the volume of fire. This was primarily an issue when troops were on the advance. During the stalemate of trench warfare in the first years of World War I, the SMLE's firepower advantage was not as noticeable. However, when German troops introduced modern mobile, small unit tactics toward the end of the war, this advantage in firepower became more obvious. This is most likely why German shock troops were usually issued submachine guns for this job, as these were better suited for this type of operation.

The SMLE action cocked on closing, unlike the Mauser design, which retracted its firing pin on opening. The Mauser also utilized forward locking lugs while the SMLE had its bolt locking lugs located to the rear. The problem with the rear locking system was that it tended to develop excessive headspace at high round counts. This could be remedied by replacing the bolt head with a slightly longer one to once again create the proper headspace for safe firing.

This was not a major issue, but the rear locking bolt of the SMLE also provided less strength overall, which is why the Mauser action tends to be more popular for custom high-powered rifles. This is especially true where very high-pressure Magnum cartridges are considered. As these bigger calibers are often needed for "safari" rifles meant for dangerous game in Africa, the Mauser 98 action dominated in this area. Another reason for the Mauser's increased popularity for use on dangerous game was its large controlled feed claw extractor. This type of extractor takes control of the cartridge case the instant it pops free of the magazine feed lips. Should the shooter panic and withdraw the bolt before locking, the round is withdrawn along with the bolt and ejected before the shooter chambers another cartridge. As this has happened on more than one occasion to both soldiers in combat and "great white hunters," the advantages of the controlled feed extractor are often appreciated.

The SMLE was not a completely new design by the time World War I began. The primary wartime model, the SMLE Mark III, had been around since 1907, and the design originated in the black powder days. The SMLE survived as the U.K.'s primary rifle until after the Korean War, when it was replaced by the L1A1 (FAL) in the mid–1950s. This long lived bolt rifle lasted as a British sniper rifle until much later, with the 7.62 NATO L42A1 variant being adopted around 1970 and lasting for another fifteen to twenty years. This isn't bad for a design that had its origins in the Lee-Metford bolt rifle of 1888.

The final bolt action design that dominated the first half of the last one hundred years of rapid technological advancement is of course the Russian 1891 Mosin-Nagant.

The Mosin-Nagant was developed at the same time as many other bolt action rifles and came about due to a severe casualty count at the hands of Ottoman Turks armed with Winchester repeaters while the opposing Russian troops had only outdated single-shot rifles. This new Russian rifle was named after its chief designers, Russian military officer Sergei Mosin and the Belgian Nagant brothers.

The design of the 1891 is similar to the Mauser in that it is a front lug locking system, but here is where the similarity ends. The 1891 uses a locking piece that is separate from the bolt body, while the Mauser bolt was a one piece design. The 1891 also does not use a controlled feed extractor but pushes the cartridge into the chamber and the extractor snaps over the cartridge rim after the bolt locks into place as with the British SMLE. Logic would suggest that this form of extractor is inferior to the controlled feed Mauser type, but both the SMLE and the Mosin-Nagant have proven themselves in combat as well as the Mauser that they fought against in not one but two world wars.

The 1891 used an interrupting feed system to prevent double feeding cartridges, which could have otherwise been a major issue due to its lack of a controlled feed extractor.

The 1891 seemed to have an overly complicated bolt design but proved rugged nonetheless and served well during the October Revolution and its following civil war. It serves to this day as a limited issue weapon in various parts of the world. The Mosin-Nagant came in several variations much like the Mauser and SMLE, but the modernized 91/30 variant is the most common. This model was the basis for the standard Soviet sniper rifle during World War II. The sniper model was fitted with a mounting block on the left side and used with either a PE (early) or PU (later models) scope. There was very little changed on the 91/30 model from the original variant. The sights on the updated variant were now graduated in meters rather than the earlier Russian imperial unit of measure, the arshin (twenty-eight inches). The sight was now protected with a circular hood. The final major alteration was the receiver's geometry, which was changed from hexagonal to circular to speed up production time and lower cost. The sniper model was similar to the standard rifle with the exception of a turned down bolt handle, which was often fitted to make the weapon easier to operate when the scope and mount were fitted.

While the 91/30 sniper model was rather crude by modern standards, it was undeniably effective, as many German troops would testify, were they still with us. Though these bolt action rifles would become the military standard for both world wars, the days of the manual repeater were numbered and, realistically, had been numbered since roughly the time of the introduction of smokeless powder. As this new powder type left very little residue behind, it lent itself perfectly to self-loading designs, which were beginning to arrive at a faster pace. The first truly successful semi-auto pistol design, the 1896 "broomhandle" Mauser (C96), is still a popular collector's piece today. This weapon not only proved effective but also had a fairly long service life. Chinese troops used it as late as the Korean War. Some were even found in enemy hands during Vietnam. This is primarily because it was an excellent design and the U.S. standard service pistol, the Beretta M9, uses a lockwork very similar in design to that of the C96.

Self-loading rifles and machine guns were already on the scene by the time of the C96's introduction. The Maxim machine gun had been designed in 1884 and was already in service before 1890.

The first self-loading service rifle, the Mexican Mondragon, was not as successful as either the C96 or the Maxim, but by the end of World War I the Browning Automatic Rifle (BAR) had arrived. While not a weapon intended to arm every trooper, it did prove the viability of a battle ready one man self-loader.

Rugged submachine guns would quickly follow, and several designs had arrived during the war. However, the M1921 Thompson would prove that automatic weapons could not only be reliable and effective but fairly compact and portable as well. Most people are familiar with the "Tommy gun" and its violin case if they have any memory of early Hollywood movies or the history of organized crime for that matter. It's true the Thompson is considered the root cause of all gun legislation, beginning in 1934. During its days of unrestrained sale, the Thompson was to acquire several villainous names, with my personal favorite being the "Chicago typewriter." The weapon's inventor was destroyed by the fact that his invention had not been accepted as the tool he intended for helping troops in the trenches and instead had become the chief enforcement device for bootleggers who were enjoying unrestrained business during this time. Judging by this, it would seem that the old adage that "God created all men, but Sam Colt made them equal" was true. With this in mind, note that the Thompson proved it made men not only equal but also more than a match for any government with which they chose to disagree. In fact, the Thompson helped organized crime enforce its will to the point that they became a force unto themselves, one with enough power and influence that the U.S. government was forced to negotiate with them at times as equal powers, though I suspect this aspect of history had more to do with money and corruption than the submachine gun.

At any rate, the self-loading firearm proved quickly that it was here to stay, though it would take some years for the world's armies to accept the self-loading rifle as a standard weapon for the individual soldier. This very process did not begin in earnest until the U.S. Army formally adopted the M1 Garand in 1936. The Army was the first major military force to make such a bold move.

Though the war began much earlier in Asia, 1939 is the year accepted by many as the beginning of World War II. As Nazi Germany invaded Poland in September of 1939, the 98K was the standard German service rifle. It would remain so for the remainder of the war.

The British, too, kept their SMLE in its most common variants, the MkIII and the newer No. 4 Mk1, which had the shorter forend with its exposed muzzle portion of barrel. While the United States had adopted the Garand in the mid–1930s, most of the world decided that the bolt action was a better choice for a standard service rifle. There were exceptions to this to some extent, however. It is well known that the Soviets had been developing semi-automatic rifles for some years.

The first attempt at a Soviet self-loading rifle was the Simonov AVS-36, which proved unsuccessful, and a newer design was developed just a few years later. The Tokarev SVT-38 was developed as a replacement for the Simonov. This new rifle was a gas-operated rifle and used the same cartridge as the standard bolt action Mosin-Nagant rifle.

The SVT-38 saw action during the war with Finland in 1939–40. It proved to have a great many issues and an improved model was quickly introduced. The improved SVT-40 was not a great deal more successful. While the SVT-40 was in fairly common use by the time Operation Barbarossa began, it never reached production numbers anywhere near those of the Mosin-Nagant or even the PPSh-41 submachine gun. This was because the SVT-40 was more complicated and time-consuming to produce. That is not to say that the SVT-40 was a complete failure. It was produced well in excess of a million units and was used by the Germans to a considerable degree. This was because large numbers had been captured during the first months of the Soviet invasion. The SVT-40 also had some influence on a German wartime semi-automatic design, the G43. This was one of a number of unique German designs that included the FG-42 and the StG 44, the world's first true assault rifle.

The G43 is of interest for the purpose of this book, as it is an early form of battle rifle like the M1 Garand and the SVT-40.

While the G43 was used to a considerable extent (almost half a million units produced by war's end), it showed up too late in the war to have a major influence on the outcome. It nonetheless had a major influence on the firepower of the German infantry squad. The lack of larger numbers of G43s available to German troops early in the war was likely the reason of making such widespread use of captured Tokarev SVT rifles. It must be remembered that there was a general shortage of German arms throughout the war. The Tokarev was not the only Soviet weapon that the Germans made widespread use of; the PPSh-41 was also popular with German troops, mainly because of its seventy-one-round drum magazine and generally better cold weather performance when compared to the standard German MP-40 9mm submachine gun.

The German army closely studied World War I data in the years after their defeat. This was true during both the Weimar republic and early Nazi rule. The motivation behind this close evaluation was no doubt to help them determine the reasons for their defeat and avoid making the same mistakes again. Obviously they had intentions in mind already. One important aspect, they concluded, was the element of armor in allowing the advance of an army during what had been largely a stalemate on the Western Front.

One particular tactical/technological breakthrough of great impact was the British introduction of the tank. More than anything, it was this invention that brought about an end to the stalemate on the Western Front. Russia and Germany fought a different war of maneuver in the East and this type of warfare was to become common in the West as well, once the tank made its combat debut.

The importance of this invention was not lost on the Germans and, during the interwar years, the German army began to utilize new technology to its fullest extent (including aircraft). These newer weapons of war being put to their full use enabled the German army to achieve some remarkable victories early on in World War II. The rapid advances allowed by these new weapons came to be called "blitzkrieg." Whether or not this was an intended tactical reformation by the German army is still open to much debate. Many historical arguments have reasonably concluded that the Germans never developed "blitzkrieg" concepts as a whole. However, some historians feel that the concept is merely an extension of the long held Prussian strategy of fighting decisive engagements to better support a rapid victory as part of a larger overall military strategy.

Regardless of the underlying reasons, an effective self-loading standard service rifle would have been a great addition to the "blitzkrieg" type of fighting initially carried out by the German army during the early years of World War II.

While battle conditions have much to do with the tactics to be applied at any given time, it became obvious after years of stalemate on the Western Front during World War I that static front lines did little other than to facilitate mounting casualty numbers. The mobile small unit tactics applied by armies later in the war helped prove the effectiveness of such tactics. This is largely why such tactics have remained in place relatively unchanged since that time. The additional effects of armor on the battlefield only helped accelerate these changing tactics. The immense distances that existed on the Eastern Front created the need for a more fluid type of fighting throughout World War I. This was the primary reason for the lack of a similar stalemate war in the East.

So due to the introduction of armor and the other new technological advances (especially aircraft), the nature of warfare had forever changed. Only with the very recent new technological introductions have we seen any deviation from this basic method of warfighting. With the introduction of new technology like drones, there may be basic strategic modifications necessary in future large scale conflicts. As only a few nations have the ability to produce this type of equipment, it will be some time before it becomes commonplace.

When World War II began in Europe, the German military was not the superadvanced and highly mechanized force as is often portrayed in historical accounts. In fact, a great many aspects of the German military were quite contemporary and in some ways the German equipment was outclassed by Soviet equipment. While Germany did lead the world in some technological areas, as a whole the German military was still lagging more than Hitler would have preferred.

Part of the reason for this was financial, as Germany had largely turned its economy around, but it had not had the time it needed to fully modernize its military to the extent Hitler had wanted. This was entirely his fault, however, due to his treatment of neighboring countries.

What is odd, though, is why the German army chose to adopt the Mauser 98K bolt action as their standard rifle in 1935. This was just prior to the U.S. and the U.S.S.R. adopting some of the first semi-auto service rifles. The decision to adopt a bolt action does not fit with a modern army looking to have the best and latest technology at its disposal. This is especially true if one considers the German military as the übermodern force that many considered them to be. In fact, there is a strong case to be made that this is proof that the German army had not developed the "blitzkrieg" approach to total war, as a self-loading rifle would be essential to supporting such a rapid form of troop advance. While Germany did have modern submachine gun designs, these were limited issue weapons, not standard service rifles. At this time, however, Germany viewed the rifle from a tactical perspective as a supporting weapon for the squad machine gun, which at the time was the MG-34 (later the MG-42). From this it was clear that the lessons of World War I had not been lost on the Germans. The basic tactics of troop maneuver made the machine gun less of an influence than it was during World War I, however. It would seem that the Germans made a basic mistake in underestimating the importance of the infantryman and his rifle in the overall approach to modern warfare. Soon after the Germans invaded the Soviet Union in June of 1941, large numbers of Soviet units were quickly encircled and taken prisoner. This was largely due to the rapid advances allowed by the modern German "blitzkrieg" tactics, though Stalin's initial disbelief, denial and inaction were also to blame here. However, as a by-product of these massive prisoner of war roundups large numbers of Soviet SVT-38 and SVT-40 semi-auto rifles were captured as well. The Germans were fairly impressed with these rifles and the advantages they brought with them on the battlefield. The Germans quickly put them to use. Far from perfect rifles, they did offer a superior advantage in firepower over the 98K in use by the German army.

As a result, the primary semi-auto rifles that had been developed by the Germans, the G41(w) and G41(m), were to be replaced by the new G43 midstream during the war. Yes, the Germans had done work on semi-auto service rifles of their own prior to the war. Neither the Mauser developed G41(m) nor the Walther designed G41(w) proved a satisfactory weapon, however, and not many were produced or issued as a result of their poor performance and more involved manufacturing process. Both designs were late to the table and this again shows that not much serious thought had gone into developing a good semi-auto German service rifle until the Nazis were in desperate need of one; by then, of course, it was too late to have any outcome on their inevitable defeat. The G43 and StG 44 may have helped prolong their defeat however.

While the Germans were busy rolling towards Moscow during the first months of Barbarossa, the captured SVTs were being examined both in the field and some, no doubt, after being sent back to Germany for further evaluation. It should come as no great surprise then that the later G43 rifle had some similarity when compared to the Tokarev designs. More specifically, the gas systems were very much alike. The gas system was one of the weaker areas in the design of the earlier G41(m) and G41(w) rifles.

Improving design in this area made the G43 a much better overall rifle than either of its early cousins. However, by the time the G43 entered mass production the tide had already turned against Germany. In all, fewer than half a million G43s would be manufactured and this just wasn't enough to alter the outcome for the Nazis. Had the G43 and, of course, the StG 44 assault rifle been standard issue prior to the start of the 1939 invasion of Poland, things may have ended differently for the Soviet Union, and maybe even for the war soon to be fought in the West for that matter. The additional firepower available to the average German infantryman when equipped with these types of rapid fire weapons would have enabled the Germans to better halt any attempts by Soviet troops to advance on a given position. Massed infantry assaults were a common Soviet tactic during the war, as Stalin cared less about high casualty counts than he did about beating the German army and helping to restore his image among his own people, which had been tarnished somewhat after his poor initial handling of the invasion. As a result, many Soviet units were pushed to advance at any cost and the lucky ones were equipped with PPSh-41 submachine guns rather than the much slower firing 1891 bolt action. The push to advance usually came from Communist Party members with machine guns behind the troops. The Germans were doing this type of motivational approach as well during the final days in Berlin. SS units were moving through Berlin and shooting suspected deserters instead of fighting the enemy. This mentality may have had much to do with their defeat, as well as poor Soviet performance in the early days, but that is another subject altogether. Suffice it to say, neither regime will be missed by any other than the like minded. At any rate, the short effective range of the Russian submachine guns helped to encourage the Soviet troops to advance and close with the enemy if they were to have any hope of survival. When talking about effective range in the case of the 7.62×25mm cartridge used in the PPSh-41, we are talking about two hundred meters at the outside, and that is pushing it. Some Soviet divisions were armed with only submachine guns to help with this aggressive tactical approach. While this may seem a cold hearted strategy by Soviet leaders, it proved effective and made Stalin happy. Often that was more important than death to some higher-ranking officers, though the Soviets did end up turning out some top commanders like Konev (defender of Moscow) and Malinovsky (master at avoiding the encirclements that befell other commanders). The seventy-one-round drum magazine of the PPSh-41 along with its full-auto capability allowed Soviet troops to keep a much better concentration of fire on German positions than would have been the case with the 91/30 Mosin-Nagant bolt action. The reliable and relatively light DP drum-fed machine guns helped here as well.

When compared to the rate of fire offered by the German five-round 98K bolt actions, the PPSh had a tremendous advantage in close range firepower. The tendency for the 98 bolt action to bind when being closed (one of the 98's few weaknesses) didn't help the German troops here, as this binding generally occurred when pushing the bolt at an angle rather than straight. It proved more of a problem when attempting to rapid fire the rifle, something for which it was not designed. Once engagement ranges opened up beyond 200–250 meters, however, things changed very much for the Russians with regard to their close range advantage.

However, if the German army had been able to arm troops, prior to Barbarossa, with the G43 and the StG 44 on a massive scale, then it becomes easy to see how any attempt by Soviet troops to take the offensive could have been far more easily slowed or halted outright. Russian casualty counts could have been double or triple the already astounding numbers that they were. It may have even depleted the East Asian troops Stalin had been holding in reserve until the 1942 general offensive.

Fortunately, Hitler was not the expert at evaluating new small arms systems and theory that he considered himself to be and he rejected the StG 44 initially. It went into production anyway against his will as the MP-43. This designation was given to help keep the weapon's issue from Hitler's notice until it had proven its value in combat, as its designers (the real experts) knew it would. Whether or not he felt the same regarding the G43 is not as clear. The G43, however, would have given the German army a rifle very similar to the U.S. M1 Garand in capability, and it appears to have been more satisfactory in service than was the SVT series used by the Soviet troops.

While the Germans had made unsuccessful attempts with the G41 series, it seems that not enough direction had been given to push for the fielding of an effective self-loading rifle early on. It also seems clear from many historical accounts that the widespread use of the Tokarev SVT series within the Soviet army was something of a surprise to many German troops involved with Operation Barbarossa at the start of the invasion. My best guess here is that far too many young German soldiers had begun to believe the propaganda coming from back home regarding the inferiority of the

Russian people. So when the Wehrmacht began to encounter such large numbers of high-powered semi-automatic rifles in Russian service, they were naturally somewhat shocked, never mind that the Tokarev wasn't really that great of a design overall. One didn't notice such things when on the receiving end.

This is not to say that Germany had simply fallen behind the other powers of the day. In point of fact, Paul Mauser had put forth designs for self-loading rifles far earlier than World War II, even World War I for that matter. The earliest Mauser patents for semi-auto rifles date back to roughly the end of the nineteenth century, roughly the same time that his 98 bolt action was entering production. This makes perfect sense considering my earlier mention of the large number of new designs coming out shortly after the adoption of smokeless powder.

Though John Browning tended to be much more open minded regarding his choice for operating method, Mauser only attempted work on recoil systems; no gas-operated designs were ever put forth by the Mauser brothers during this time. To be sure, recoil operation does tend to allow for greater degrees of fouling without adversely affecting performance, so his focus here does have some solid logic behind it.

As the Germans definitely stepped up their efforts after the war had gained momentum, it is obvious that they well understood the effects that semi-auto rifles had on the battlefield. As mentioned, prior to the war's beginning and even during 1940–42 the Germans held the tactical belief that the individual rifleman should support the machine gun crew. Again, this appears to be a remnant of the memory of World War I. The modern tactics that the Germans were largely responsible for relied on the machine gun being the supporting weapon for the rest of the squad. This oversight never made sense coming from such detailed tacticians. The huge number of casualties caused by machine gun fire during World War I forced even the top military skeptics and "old schoolers" to face the reality that the belt-fed weapon was not merely a passing fad; a new form of death had been brought to the battlefield and tactics and even overall strategy would have to be vastly different from that point on. It had become a key weapon in modern warfare. Apparently, the Germans considered it "the" key weapon regarding infantry tactics. This was an example of the Teutonic tendency to take a sound idea and go too far with it. The real cause of why men die in war is the war itself; death can come from any number of devices. Heavy artillery fire and aerial bombardment have been known to cause quite a few casualties in their own right. Regardless, it would seem that the German oversight regarding early semi-auto rifle development was rooted more in the tendency for an army to always fight the last war in formulating plans for the current one. This tactical oversight seems rather strange coming from the country most often credited as being responsible for the development of modern mobile small unit tactics.

The primary weapon carried by the shock troops (the original stormtroopers) first used in the later years of World War I was the then new Bergmann MP18 9mm submachine gun. The light machine guns like the Madsen and Bergmann MG15 were not really "light" enough for storming trenches in a heartbeat. The Bergmann weighed nearly thirty pounds empty and the captured but reliable Madsen often used by German troops was still twenty pounds in the same condition. The need for something lighter was clear. While the Madsen was probably the first truly successful light machine gun, it was not an official German weapon. Though a weapon such as this or the even lighter U.S. BAR would have been nearly ideal for such small unit tactics, not enough of these were in use by the war's end to allow these weapons to be considered standard equipment for such troops. While these weapons were relatively new, they would have a lasting influence on modern tactics and would lead to the general purpose machine gun (GPMG) concept in later years. This was because the heavy, liquid-cooled machine gun that was the World War I standard proved too cumbersome for mobile infantry, which would be the norm from the end of World War I up to today. It is most likely the Lewis light machine gun that had the greatest influence in this regard, however, as this proved the remarkable versatility of a "light" automatic weapon.

While this discussion of light machine guns may seem off topic, there is a direct relation to the intended purpose of both this type of small arm and the semi-auto service rifle. Both the machine gun and the semi-auto rifle are intended to provide high rates of accurate fire. The only difference in intended use is that machine guns are suppressive in the nature of the fire they provide, where the rifle is less useful in this purpose merely because of the rate of fire of which it is capable. Remember that the first machine guns in service were not overly portable and were set up as defensive weapons to prevent effective infantry advance. The fact that both sides employed them is largely what led to the stalemate on the Western Front (limited room for movement was also a major factor). Neither side could effectively advance.

Light machine guns were developed shortly after the Maxim's introduction, as it soon became obvious that a portable variation would be quite useful in allowing troops to attempt any forward movement.

The suppressive nature of machine gun fire gave the troops that split second needed to get moving from one position to the next. I am speaking of course of covering fire being provided during an advance. The aimed automatic fire helped keep enemy heads down just that much longer and aided the troops in their movement.

What the semi-auto rifle did was give the individual this same capability, though to a much lesser extent. This is very much an extension of the advantage enjoyed by the ten-round SMLE British bolt action over the M98 five-round bolt gun (with both designs being standard for a second stage appearance).

When the Germans began Operation Barbarossa in the early morning hours of June 22, 1941, the initial Soviet response was (nearly) non-existent. Stalin had even issued orders not to fire unless in direct defense against an attack. As a result of both the rapid early German advance and the initial Soviet indecision, many Soviet units stationed near the border areas were quickly cut off, encircled and captured along with huge stores of Soviet arms and equipment, some of the most valuable being the armor, artillery and fuel supplies, as much of the newest equipment was there as a deterrent. There were exceptions to the initial Soviet lack of effective defense, as at Brest, which held out for some time, but by and large the Germans advanced far faster than even some of their own commanders had expected.

Due to the severe early German advance in June and July, the huge quantity of captured Soviet equipment was pressed into German service. This included large numbers of SVT rifles, as mentioned. There were several factors behind this use of foreign gear. One was the German surprise at the Tokarev rifle's capability in terms of firepower, especially when compared to the standard German 98 Mauser. The Germans, like most soldiers, quickly learned respect and admiration for effective tools of war. While the SVT-38 and later SVT-40 should not be considered battle perfect designs and were not quite up to the standard of the M1 Garand, they were far better than the G41 series developed by Germany. Another valid reason for German troops using captured Soviet weapons had to do with the rapid advance of the German army during the first months of the invasion. Supply lines and logistics had been strained to the point of collapse. As the German troops had already captured large quantities of SVT rifles and vast amounts of ammunition, it made perfect sense to use weapons that were capable of chambering locally available ammunition once the original captured stores ran out. This was also helpful to German finances as well.

In fact, the Tokarev had become so common in German hands on the Eastern Front that the German army even issued an operator's manual for the weapon. It was also given a German designation, the G.259(r) (SVT-40). The SVT-38 was given the designation of G.258(r). Oddly, it seemed to be more highly thought of by the Germans than by the Soviet troops to whom it was issued. Many consider this to be because it was seen as too complicated a design for maintenance purposes, especially when compared to the more common 91/30 Mosin-Nagant bolt action. It was certainly more costly and time consuming to manufacture. As a result, the total numbers produced of both SVT models was far less than the number of 91/30s made during the war. This was crucial to the Russians, who were hanging on by a thread at one point and were in desperate need of rifles for issue to their large army. Why waste time making one when the factory could almost get two out the door if the simpler model was produced? There is also the fact that the Tokarev designs were not as durable as the Mosin-Nagant and the Soviet troops had a tendency to provide a lesser degree of maintenance for their weapons. This last aspect is largely dependent on the individual soldier, however, and there were no doubt exceptions. It is generally regarded that Soviet troops expected maximum performance from their arms concerning reliability. Many troops tended to view the simplest, most rugged designs with greater admiration. When taken as a whole, this helps explain the popularity of the PPSh-41/PPS-43, Degtyaryov DP series, and the SG-43 Goryunov belt-fed medium machine gun. These were all very successful and simple Soviet designs that would help win the war for the Russians.

The Soviets were also better able to keep their offensives going in wintertime, as they were better prepared than the Wehrmacht for cold weather warfare. However, as a result their weapons tended to receive harder use. The Russians expected arms to be able to withstand such conditions without failure to perform. The Germans quickly realized this as well, as they were forced to defend against the Russian onslaught after only being on the offensive for a short period of time relative to the entire war. This is why the PPSh-41 is also so often seen in the hands of German troops in war photos. This submachine gun was often chosen in preference to the Germans' own MP-40 9mm.

It is quite likely that the Wehrmacht troops used Soviet weapons as a matter of convenience during the opening months of Barbarossa, but as time passed, the Germans came to respect the Russian designs more and more for their quality of design and their performance under harsh combat conditions. This admiration only increased as the German troops became more familiar with the workings of the various Russian models.

Another possible factor is that as the Germans came to meet stiffer resistance in the fall of 1941 they began to realize that the Soviet troops were not the inferior *Untermenschen* that they had been led to believe. The Russian trooper was more often than not a hard charging, tough peasant soldier who could withstand the harshest of combat conditions. The Soviet soldier was also used to getting the job done with the least amount of specialized equipment possible. He was able to make do with far less than what the average German trooper would consider minimal equipment. This is not to say that the Soviet soldier was a completely neglected individual. Not only was some of the Soviet equipment equal to what the Germans had at their disposal, but in many cases the gear was superior. The Soviet winter uniform was a good example. The efficiency of the Soviet winter clothing enabled the U.S.S.R. to carry on with operations while the German troops were losing extremities to frostbite, though much of the situation faced by German troops during the early winters had more to do with logistics and planning. Hitler had assumed that Barbarossa would bring about the collapse of the Soviet Union within weeks. Given the speed of the initial German advance, he was nearly proven correct. However, this led to no winter gear being issued to the troops at the appropriate time (prior to the onset of winter). This makes sense, as Barbarossa began in the early summer, but by the time the advance had stalled with the Moscow skyline visible to the forward most German units an early Russian winter had hit and hit hard (one of the coldest on record, oddly). Even by this time, many German units had no winter gear. The length of the logistical train was almost hard to imagine, and this no doubt had some major influence on the lack of proper German winter gear.

However, my thinking here is that the better preparation for winter fighting by Soviet troops may have led some German soldiers to the wrong conclusions, once again taking a sound idea one step too far. The better winter war experience of the Russians may have led some German troops to conclude that the weapons themselves were superior when they were merely better winterized. In a sense, this is a superiority of sorts. The simple Russian designs were perfectly suited for cold weather warfare, as they were designed this way. Russian weapon designers were used to native troops having to operate in cold weather for a good part of the year, and this need was always put forth when considering a new design for adoption. Also, the Russian experience with cold weather warfare likely left them better adapted to caring for their weapons under such conditions. The recent beating taken by the Finns in the winter war in 1939–40 also made them realize the importance of caring for weapons in extreme cold.

In any case, the Germans had come to closely study the Russian designs and later emulate them in addition to making good use of the captured stores.

By the time the G43 began to roll off the production line in late 1943, the Germans had already lost the war, whether they chose to see it or not. Stalingrad had already ended the great German 6th Army and Hitler's dreams of conquest had been shattered, though he may not have yet seen it. With the ending of the Battle of Kursk, the final German defeat was only a matter of time.

Despite this, the G43 was to continue in production until the war's end. The respect that the Germans had gained for the Tokarev SVT series had led to the similar gas systems between the two. While the G43 was not a particularly well-finished rifle, it proved effective in combat just the same. This poor level of finish is due to its introduction after the war had already turned against Germany. This was fairly common with many wartime small arms, but it was especially so with late German war production.

While the Walther G41(w) proved to be a disappointment in the field, the primary problem seemed to lie with the design of the "Bang" type gas system. This method differed from the traditional gas operation method, which used a small hole tapped in the barrel near the muzzle.

In the Bang (the Danish designer's last name) gas system, a muzzle cup was driven forward by the exiting gases and this motion was used to operate the mechanism. The system was similar to the gas trap system tested on early M1 Garand models and it worked no better for the G41(w) than it did for the M1. It was a heavy, bulky and complicated system that fouled easily. This is odd, because the Germans specifically asked for no hole tapped gas designs for fear of fouling issues, though many successful gas-operated designs used the tapped barrel method as a means of gas delivery to a piston or some other mechanism.

As the SVT Tokarev design had so impressed the Germans, it was found that using the Tokarev based short-stroke gas system when mated to the basic G41 receiver system created a usable rifle, and the G43 was born.

In addition to the new gas system, a detachable magazine was added to help create a much better overall weapon. On the G41(w), a fixed magazine was used for fear of troops losing a detachable magazine in combat. This serves to prove that even technically capable people can still be prone to backward thinking at times. Apparently, the idea of spare magazines and rapid reloading capability didn't click in someone's head at the time.

While the G43 entered service late in the war, it saw action on all fronts and stayed in service among East German and Czech troops for several years after the war's end.

Though the scope of this book is the battle rifle, the small arms that came before it were highly influential to its development and the world's first true assault rifle was definitely a close relative. This is especially true when considering the selective fire capability of both (though some battle rifles were modified for semi-auto fire only). The StG 44 took a more practical approach to effective use in rifle design than many had realized at first.

The limited range of this weapon was not seen as an issue for the majority of combat operations. It also allowed savings in raw materials needed for production of rifles. In addition, the intermediate cartridge design used by the StG 44 offered a similar savings where ammunition production was concerned. After various nations had studied the design following Germany's defeat, the logic behind the design was clear. This was to have a heavy influence on later designs from the U.K., in the form of the EM-2 "bullpup." The FN FAL was also heavily based on the capabilities of the StG 44. Most important, the Soviets themselves greatly appreciated the capability of the German "storm rifle," as the AK-47 very closely follows its design in both layout and intended purpose.

The Germans had almost created the ideal weapon for their needs with the StG 44, and again, had the rifle been available at the start of the war things could very easily turned out very different. The large number of Soviet PPSh-41s encountered by the invading German troops was a shocking wake-up call that the Mauser 98K bolt action was nowhere near as effective for close combat as was the Russian submachine gun and that they would need to have something to counter the small Russian submachine gun. The MP-40 had been in service since Operation Barbarossa began, but while adequate, it was never known for being the most reliable submachine gun in service.

The PPSh-41, however, was a design inspired by Soviet experience in the Winter War against Finland in 1939 and the Finnish troops were often equipped with a 9×19mm submachine gun known as the Suomi, which has long been known as being one of the most reliable submachine guns of the interwar period. The Suomi used a large capacity drum magazine, which would be directly copied by the Russians in designing the PPSh. The actual PPSh itself was an improvement over an earlier Russian model, the PPD-40. This weapon proved less than ideal and the PPSh-41 was meant to correct the flaws of the earlier model. This it did in spades. It proved cheaper to produce and was reliable to the extreme, even in cold weather, which was a basic requirement for any Russian small arm design. The Russians learned a healthy respect for the high capacity drum magazine of the Finnish Suomi, and the seventy-one-round drum magazine was one of the PPSh's best features for close combat. The Germans also quickly learned a healthy respect for this weapon and it was one of the favorite captured weapons for German troops, along with the Tokarev SVT-40.

With the proven capability of the PPSh-41 in close combat, the Germans looked at the most promising design they had, and it would become the MP-43/MP-44/StG 44. There was no difference between these guns other than the designation. Nazi Germany was never known for a simple and consistent small arm categorical system. The new StG 44 proved to be just what the German troops needed. It was capable of returning full-auto fire and was far more powerful than the Russian submachine gun. It proved reliable in the Russian climate, even in winter. Its primary drawback was that it was rather heavy at ten pounds empty. The Germans were willing to live with this in order to counter the Russian threat. The StG 44 was mass produced as quickly as the German war machine would allow, but by 1944 the Allied bombing campaign was beginning to take its toll and Germany was continually short of pretty much everything. By war's end, less than half a million StG 44s had been manufactured. This was roughly the same number as the G43 semi-auto rifle that was also in high demand and short supply. While these weapons did not turn the tide of the war for Germany, both had a profound influence on the later design of the battle rifles that would follow.

If the assault rifle was clearly the infantry rifle concept of the future, then why the battle rifle at all? The battle rifle would seem to be a step in the wrong direction if that were the case. The answer has already been mentioned to some extent earlier in this book. Following the war's end and for many years after, the U.S. was by far the dominant economic power in the world. Thus the top U.S. commanders were able to use this position to influence final decisions when it came to arms adoption. Remember that the FAL was initially built to utilize the 7.92×33mm "kurz" cartridge, the same round used in the StG 44. The 7.62mm NATO FAL was the one adopted.

The StG 45 that formed the basis for the CETME rifle and later G3 was also designed to use the 7.92mm "kurz" round. Both the G3 and CETME were adopted in 7.62mm NATO caliber rather than the much smaller, cheaper and easier to use "kurz" cartridge. Why would the U.S. push to insist on these caliber restrictions when combat studies dating back to World War I showed 7mm bullet diameters to be the most effective for lethality? Later German studies showed most World War II combat taking place at ranges of three hundred meters or less. The answer most likely lies with what can, in some terms, be considered the first battle rifle, the U.S. M1 Garand.

Officially adopted in 1936, the M1 Garand was modified to handle a .276 Pedersen cartridge, which was looking like the new U.S. service rifle round at the time. For several years during the Garand's development, this .276 round was proving to be a good performer. Then, just prior to the Garand's adoption, General MacArthur insisted on going to the .30-06 due to large quantities of existing ammunition stores.

The first Garand prototype showed up in the early 1920s. A detachable magazine would have allowed for a simpler and lighter design with better potential firepower, but the U.S. Ordnance requirement called for an internal magazine that used charging clips of some type. It was felt that detachable magazines would be too easily lost, rendering the rifle a single loader. Apparently, the Americans saw things the same way as the Germans in this regard.

Garand's working model introduced the familiar eight-round "en-bloc" clip and its accompanying distinctive "ping" sound that it made when ejecting from the rifle when emptied. The initial adopted model also used a "gas trap" method of operation that utilized an attachment directly ahead of the barrel's muzzle. The device was shaped and spaced to allow the proper volume of operating gas to be diverted to a chamber located beneath the barrel. This gas then pushed against the piston to operate the mechanism. This system was to be standard for the first years of production. However, in 1939 it was decided to switch over to the simpler gas port system that we know today. M1s in service were retrofitted with the new gas system. This would be the M1's design for the remainder of its production life. There were to be minor design alterations in coming years, but these were primarily for improving the service life of some highly stressed components and to help with lowering production costs or reducing the manufacturing time required.

While the U.S. Marine Corps did not initially adopt the M1 along with the Army, this would change in 1940 after the Corps had tested several other designs, like the Johnson M1941 recoil-operated .30-06 and its bulky ten-round built in rotary magazine.

After testing under combat conditions, it appeared that some liked the Johnson and its slightly softer kick. This was not to happen. Largely due to the amount of time and money invested in the M1, it was decided not to adopt the Johnson. Some testing done by the Marine Corps actually concluded that the Johnson was the better rifle. The conflicting test results make one wonder how valid the tests were at that point, due to the interest in the M1 by that time. It did have some issues with its recoil operation and bayonet usage. It could not reliably fire with a bayonet attached due to the added weight on the barrel, which changed the recoil forces available to operate the mechanism.

In the end, the Garand wound up being the standard for both the Army and the Marines. This was not necessarily a bad thing, as the M1 gave top notch performance at a time when a good semi-auto was a distinct advantage for U.S. troops.

The design of the M1 involved a long process, one of continual improvement. This was true even after production had begun, starting with the gas system overhaul. While the "en-bloc" clip was not without its flaws, it did work well enough. The design specifications necessary for the feed system created unwanted weight and feed complications for the rifle. This added complexity was needed, as moving parts had to be designed to create the upward feed pressure and to alternate or distribute this pressure between cartridges on both the left and right side of the clip as it held the rounds in a staggered configuration. The feed mechanism also was responsible for throwing the clip clear of the weapon once empty. All of these feed issues are handily taken care of in a detachable box magazine design. The stepped follower typical of the staggered box rifle magazine places the cartridges in proper alignment and the spring acting against the entire bottom of the follower handles the rest. This is a far simpler solution to a feed system, which is why it is still so popular today. While box magazines are not as rugged as the internal and well-protected M1 feed system, box magazines are cheap and easily replaceable. To be sure, they are not as low cost as the stamped spring steel clips used by the M1.

The only real disadvantage to the M1's feed system was the skill needed to load a clip through the top of the M1's receiver without hurting one's thumb. In order to load the M1, it was necessary to lock the action open. This was the easy part. All that was needed was to pull the operating handle fully to the rear and it locked open automatically, provided the rifle was empty. To insert a loaded clip, the small raised studs on the clip had to be properly lined up with the corresponding notches in the rifle's feed well. In use, this was fairly easy to do, as the user quickly learns to press the base of the clip against the rear of the receiver well, which was an instinctive movement. The real trick came when it was necessary to press home the clip in order to release the hold-open lock. The clip engages easily enough, but the bolt often (usually) hangs up before stripping the first round from the clip and running it into the chamber. Again, this is not a bad thing, as it was often necessary to smartly rap the rear of the cocking handle forward to chamber the first round. This bolt hang-up ensured that the user's hand was out of the way when the bolt ran home. This is a case of sluggish operation preventing possible injury. If the bolt and carrier do not hang and instead run home, the result is that the shooter's thumb is in the path of the bolt and is often pinched. This is usually the case if the shooter is too slow in moving his thumb out of the way. This condition is often called "M1 thumb."

The Garand was far from perfect, but it did have some excellent qualities as well. Its accuracy has been debatable over the years, but it was clearly good enough to warrant the M1's adoption as a standard issue sniper rifle with minor modifications. Two basic sniper models were adopted, the M1C and M1D. The primary difference between the two was in the scope mount design and flash hider pattern. The M1D had a simpler scope mount and was often encountered with a cone type flash hider. The M1C had a more involved scope mount and the flash hider was a prong pattern that screwed on in place of the gas cylinder lock. While this flash hider design usually proves most effective, its open ended prongs tend to snag foliage and everything else. Personal experience regarding M1 accuracy has been outstanding, equal to that of many better bolt action designs. Group sizes are easily sub-minute-of-angle (MOA), and my own M1 is capable of roughly one-half MOA with good handloads.

The standard scope used on the M1C and M1D was usually an M84 (M81 and M82 were used early on). The M84 used a low power magnification of slightly more than 2×. The scope mount held the scope in an offset position to the left of the receiver to allow loading of clips and to stay clear of ejecting cases, which tend to go up and back to the right. To help with the offset scope, a leather cheek piece is often fitted to the sniper models to help maintain scope/eye alignment and a proper cheek weld with the stock. This leather piece is laced on and cinched tight. Lyman was the maker of the early M81 and M82 scopes and these were primarily their Alaskan commercial model. The M84 was made by a different company named Libbey-Owens-Ford, and while these scopes were relatively basic and limited, especially by today's standards for sniper scopes, they did the job just the same.

The gas system of the M1 was a long-stroke design. The M14 and M1 carbine all have a unique design characteristic. On all three the cocking handle also acts as the bolt carrier, and in the case of the M1 Garand the gas piston as well. This helps simplify the design but also allows the cocking handle to act as a forward assist on the rare occasion that the bolt doesn't lock completely on carrier return.

When the M1 design was changed to use the conventional gas port in the barrel, the hole was located near the muzzle to help reduce pressure on the operating rod and its related components. This helped to ensure that the bullet had left the barrel and the residual gas pressure had dropped to a safe level before bolt unlocking could occur.

The gas cylinder on the M1 was constructed from stainless steel to better keep corrosion at bay. The high heat and use of corrosive primers in military ammunition tended to help speed up the corrosion issue, and stainless steel is just better at dealing with this. The problem with the use of stainless steel here is that it is a difficult metal to coat with any traditional non-reflective covering. This component was treated to allow it to hold a non-reflective coating, which tended to wear off rather easily, and troops had to learn ways to deal with this. These field expedient measures would work well but would quickly wear off and the process had to be repeated often.

Fortunately, this usually only involved use of a carbon rich flame or even a match if that was all that was available. This was necessary, as a shiny exterior part is not a good thing to have on a service rifle meant to keep a low profile.

Given that the M1 Garand was the first U.S. attempt at a standard issue self-loading rifle, it gave stellar service throughout the war. It is well known that General Patton was quoted as saying that the M1 was "the greatest battle implement ever devised." The M1 proved reliable in all theaters of the war. It saw use from the start of U.S. involvement in North Africa and was used throughout the Pacific from 1941 to the war's end. It was carried by troops during the Normandy invasion and on into Germany until the day of their surrender. If there had been any doubts regarding the self-loading rifle as a viable service weapon they were fully removed by the end of World War II.

While the M1 and, to a lesser degree, the Tokarev SVT-40 and G43 may have been the first successful general issue service rifles, there was a much earlier design that is an even closer relative of the classic battle rifle, the Browning Automatic Rifle or BAR.

When the U.S. entered World War I in 1917, there were very few machine guns in U.S. service and there was no such thing as a self-loading combat rifle capable of being fired by one man.

As per his usual, John Moses Browning was one step ahead of the rest. By 1917, Browning had some new designs ready for production. One was the famous M1917 machine gun. This belt-fed, water cooled .30-06 weapon was America's answer to the Maxim machine gun. In fact, the M1917 is generally considered a more reliable machine gun than the Maxim and its derivatives like the British Vickers. This is something that many British World War I veterans might have argued against, as quite a few held the Vickers in high regard and considered it the ultimate in machine gun design.

In either case, these belt-fed weapons may have dominated the battlefield in 1918, but they could hardly be considered portable, one man weapons, and by no means was anyone going to be shouldering one for firing purposes.

The second of Browning's designs was to prove equally successful and long lived. This new design was very close in layout to the battle rifle that would come much later, though its intended purpose was different. The BAR would receive the designation M1918 and would become an American icon. The BAR was put into service during the last months of the war and did not see a great deal of action for several reasons. By the time that the armistice was signed in late 1918, fewer than one hundred thousand had been put into service, though this number would increase greatly in coming years.

Another reason for the BAR's limited use during the war was that the Army was afraid it might be captured and illegally copied by the Germans.

The concept behind the development of the BAR was that of "walking fire." The user was supposed to advance and fire the weapon every time the same foot stepped forward. Apparently, this was a French idea, and the one who thought it up must have been a simpleton, to say the least. A grunt should fire his weapon as needed. Anything else is a waste of time and ammo. Even suppressive fire is best limited to controlled bursts only when movement is detected.

The original M1918 weighed around sixteen pounds empty and used a twenty-round detachable box magazine (as did all subsequent models). This first model did not utilize a bipod. As a result, it was rather difficult to fire accurately in the automatic mode. The M1918 and M1918A1 were selective-fire weapons where the M1918A2 was capable of automatic fire only and was fitted with a rate reducing mechanism to offer two rates of fire, 350 rpm and 550 rpm. The earlier versions fired at 500 rpm only when in the automatic mode.

The BAR uses a conventional under barrel gas piston system. The gas system has an adjustable port and could be varied for use depending on the conditions of use. The BAR was a technically a striker fired weapon and did not utilize a conventional hammer as many know it. As it fired from an open bolt, this actually helped to keep the design simpler than would have otherwise been the case.

The BAR did have a hammer, but it only struck the firing pin after the bolt had closed and the slide had completed its forward travel. The firing pin is withdrawn and blocked from striking a cartridge case once the bolt is unlocked. The bolt locking is accomplished by camming the bolt upward into a locking recess, which is the explanation for the large hump at the rear upper portion of the receiver. The bolt is connected to the slide (carrier) by a link pin, much as in the barrel link recoil system of Browning's earlier M1911 pistol design. While the original M1918 and M1918A1 were selective fire, the M1918A2 had no provision for semi-auto and was fitted with a complicated rate reducing mechanism that held the slide back until an activator built into the buttstock had time to completely cycle through its full recoil motion. This actuator then tripped a sear lock and the sear was then free to move out of engagement with the slide.

While the BAR had no provision for a quick change barrel, it filled the role of standard squad automatic weapon through World War II and Korea.

The BAR's real contribution to small arms technology was its introduction at such an early date. Few designs in this weight and size class had even been thought of, let alone developed to such a successful level.

After witnessing its capability, the French military requested the BAR for their own troops. This was of course after the honor of providing U.S. troops with the Chauchat light machine gun upon arrival in Europe.

Other light machine guns like the Chauchat had been developed. The Danish Madsen was another early design but like most was considerably heavier than the BAR. The BAR not only weighed less than its competition; it also used a rifle type layout with which troops were naturally comfortable.

The BAR has been difficult to classify as a weapon as it was not really a service rifle, yet it did not possess the usual characteristics of a light machine gun. In the end its name was perhaps the best description for the weapon. It was an automatic rifle. Its significance in the development of the battle rifle is that the U.S. had considerable success with the BAR and perhaps the insistence on keeping the full-auto feature as a requirement of the M14 may have been due to the desire to maintain the capability of the BAR. This was in spite of all evidence supporting the fact that a nine-pound full-power rifle cannot be easily controlled when fired in the automatic mode. This the M14 proved by being nearly uncontrollable when fired in full-auto.

The hinged locking system of the BAR may have seemed rather complicated at the time of introduction, but it proved quite effective and durable and the later FN MAG machine gun used a similar locking system in an inverted position. The MAG has proven to be one of, if not the, best belt-fed machine guns ever.

As mentioned, several countries introduced their own variants of the BAR and some of these came very close to being supreme light machine guns. FN produced its own version known as the Mle 1930 and a later, improved variant, the Mle D. This second model had a quick change barrel feature, which made it an excellent squad automatic weapon, almost the equal of the British Bren. The only real disadvantage was the bottom loading magazine of the Mle D, which tended to be slower to reload when compared to the top loading Bren magazine. This FN gun was available for a time in 7.62mm NATO as well as the .30-06.

While the BAR never really fit into any one category perfectly, it ended up serving the U.S. into Vietnam and can still be found in service in some parts of the world to this day. This is largely because the U.S. gave away or sold many after it was declared obsolete. It is still considered an iconic weapon by some within the military, though most of the troopers who have used the weapon are now retired.

The BAR never became the ideal squad automatic weapon that it should have been while in U.S. service, but it served admirably during World War II and into Korea as well as the early years of Vietnam. It will forever be remembered as a classic squad automatic weapon. Though it had many faults in this role, reliability was not one of them. Due to its intended purpose of providing an infantry squad with portable, automatic fire, it is clearly a precursor to the battle rifle. It is likely due to the BAR's well-earned reputation that it heavily influenced U.S. commanders who insisted that the M1 Garand's replacement have full-auto capability.

World War II is often considered the most horrific time in human history and the number of rounds of ammunition produced during this conflict is too great to fathom, especially when artillery and air munitions are included in these numbers. However, many of the standard calibers used during that war are still standard military calibers today. Some, like the 9×19mm Parabellum and .45 ACP, remained standard until after the turn of the twentieth century, with the .45 again gaining in popularity in recent years. While the .30-06 Springfield cartridge was a standard U.S. caliber well into the Vietnam War era, it was quickly losing ground to its younger brother, the post-war 7.62mm NATO. The 7.62×51mm was not only the standard NATO caliber for much of the Cold War era, but it had already gained a good reputation for its excellent accuracy and consequently became a popular sniper caliber. Due to this military standardization (in addition to its other qualities), its commercial equivalent, the .308 Winchester, became an extremely popular round in its own right, and the .308 is still popular today for both shooting matches and hunting.

Some may wonder why the move to a second .30 caliber cartridge was necessary if the .30-06 was such a star performer. The answer is the desire to shorten and lighten the next service rifle, which was intended to be a selective-fire weapon. The closest thing the U.S. had at the time to such a rifle was the BAR. However, the BAR was far too heavy and awkward to become the standard issue service rifle. It weighed over sixteen pounds in its lightest configuration, and this was in the empty condition. Also, a shorter round allowed for a far shorter receiver length and this was one of the complaints regarding the M1 Garand rifle, which had been the U.S. Army standard since 1936.[13] A very long action was required to feed the equally long .30-06 cartridge.

The actual performance of the 7.62mm NATO differed very little from that of its older brother. Both used a standard bullet weight of 150 grains and the bullets were the same diameter, as the U.S. would not accept any cartridge with a caliber of less than .30 at the time of the 7.62mm's adoption.

Initial muzzle velocity for both rounds was roughly the same, with the .30-06 being about 100 feet per second faster.

When fired from the shorter barrels, the 7.62mm NATO loses roughly 30 feet per second for each inch of barrel removed down to a length of around sixteen inches (200–300 difference from twenty-two to sixteen inches). The M14 rifle uses a standard barrel length of twenty-two inches. Current short barrel versions of the M14 are more convenient in use yet possess more than enough power. Early attempts were made at making a short barrel M1 Garand but were not as successful as with the M14. This was most likely due to the different gas system designs.

Years ago (in roughly 1964), a special 7.62mm NATO match grade cartridge was developed using a heavier 173 grain bullet and was known as the M118.[14] This round was used by snipers on more than one occasion, but today other sniper rounds are popular such as the Federal 168 grain open tip match (OTM) bullet. An improved match grade cartridge is in use today by the United States and is known as the M118LR.[15] This cartridge uses a 175 grain open tip match grade bullet. These rounds provide greater wind resistance than the standard M80 ball round, yet they have a steeper trajectory due primarily to the lower initial velocity when they leave the barrel. These heavier bullets, however, lose velocity more slowly due to their greater mass.

The .30-06 Springfield was most commonly used in its 150 grain M2 ball configuration. The original 7.62mm NATO cartridge was known as the M59 ball round and used a similar bullet weight, but this has been replaced by the current M80 ball cartridge. There's very little difference between the two rounds other than a slight variation in bullet weight and construction, with the M80 bullet weighing about 147 grains, while performance is similar for both cartridges.[16] Ball and match cartridges were not the only loads made for these calibers. Armor piercing (AP) cartridges have been available for many years and the current versions usually use a bullet designed with a tungsten core and a lighter overall weight than their ball counterparts. In addition, .30-06 armor piercing incendiary (API) ammunition was sometimes used against targets that were easily ignited, such as fuel dumps and munitions stores. Tracers have been used on occasion, but infantrymen quickly came to realize that if they can see where the bullet is going the enemy can see where it's coming from. More specialized loads are also available for certain operational requirements. Dummy rounds and blanks are used for training, and specialized blank loadings are used for launching grenades off the muzzles of rifles, though newer rifle grenade designs can use ball ammunition for this purpose. The newer grenade designs actually trap the bullet as it leaves the barrel. While all of these different loadings are offered, ball ammunition is the most commonly issued, along with the new tungsten cored armor piercing rounds, which are becoming increasingly useful due to the move towards more urbanized fighting conditions. The added penetration of the tungsten cored rounds proves quite useful in the penetration of motor vehicles, especially those being driven by terrorists when they have loaded explosives throughout the vehicle. It is completely reasonable to assume that the armor piercing rounds will see continued service alongside ball ammunition for some years to come. The current 7.62mm NATO U.S. standard armor piercing round is the M993.[17] This cartridge uses a bullet weight of roughly 127 grains and leaves a 22" barrel with an initial muzzle velocity of just under 3,000 feet per second. In comparison, the 7.62mm NATO M80 ball cartridge leaves the barrel of the U.S. M14 rifle with an initial muzzle velocity of approximately 2,750 feet per second. This gives a claimed effective range of 800m. This is where we run into trouble. Effective range of a rifle cartridge and the rifle firing it can mean many things. First, it needs to be clearly defined what is meant by "effective range."

I remember reading an article written years ago by a World War II veteran that described a Pacific Theater sniper team laying down shooting mats and setting up a position to fire on a Japanese machine gun nest. The nest was at a distance of nearly 1,200 yards according to the spotter of the two-man team. The sniper adjusted his Weaver scope on his M1903A4 Springfield rifle and when ready he fired. The first round hit several feet to the left of the small opening of the pillbox where the machine gun was located. The sniper adjusted the scope and fired again. The machine gun fire stopped momentarily, then started yet again. The sniper fired a third round; the fire again stopped briefly, then began once more. A fourth shot cut loose from the Springfield. The machine gun fire stopped and this time remained silent. The Japanese unit manning the pillbox apparently got the message. Each man put behind the machine gun would likely die from a gunshot wound.

That account probably illustrates the true capabilities of a sniper team better than anything else. The range at which they were firing was well beyond the often cited maximum effective range of the M1903 series, which is generally regarded to be 800 yards. This happens to be the maximum distance at which the rear sight can be set on the M1903A3 variation, though earlier versions had longer range sights installed. These earlier versions could actually be adjusted to over 2,800 yards. It is doubtful that point targets (individuals) were often hit at this range.

To further complicate matters, iron sights begin to lose the ability to engage a human target beyond 500 yards, limiting their effective range against point targets, though area fire with iron sights may be effective well beyond this distance. This is because a human sized target becomes so small beyond this range that the average human eye cannot effectively hold the front sight blade on it. The shooter personally has more to do with the rifle's effective range than any other factor, within the limitations of the cartridge, that is. One person may be able to consistently hit a target at over twice the range of another shooter. There are many factors involved in this equation, too many, in fact, to be covered entirely. Some major factors, however, are the overall skill of the shooter, familiarity with the rifle being used and the user's knowledge of the round's performance at various ranges, the zero distance of the particular weapon, and current environmental and combat conditions, both of which tend to be highly variable. Of course the shooter's eyesight is a major factor as well.

Optical sights greatly improve a rifle's effective range, especially when using longer range cartridges like the 7.62 NATO. This is definitely true when using magnified optics and if those optics are being employed at ranges beyond 500 yards.

The difference in performance between the 7.62mm NATO and the .30-06, as mentioned before, is minimal. Both of these rounds have been used over the years to successfully engage point targets at ranges of over 1000m. The record sniper shot for the 7.62mm NATO is actually 1,367 yards.[18] This was made during the recent war in Iraq. Given this record, this distance is clearly within the caliber's effective range, as the target is most likely not with us anymore. Some individuals (like the shooter who performed this feat) just tend to act in a hostile fashion when their fellow service members are being shot at; go figure.

While this is an extreme example of a cartridge's effective range, other shots have been performed with this caliber at similar ranges. It should be remembered that this was over one and one-half times the normally accepted effective range of the M14 rifle. A fairly standard rule for determining the effective range of a rifle and cartridge combination is the maximum distance at which the average skilled shooter can reliably hit a point target, reliably meaning better than 50 percent of the time. Again it must be considered that a highly skilled shooter will be able to extend this range considerably.

To better illustrate the difference between a battle rifle and an assault rifle in terms of their respective effective ranges, let's compare the 7.62mm NATO to the 5.56mm NATO, which is the ultimate purpose of this book anyway. Iron sight limitations aside, let's say the effective point target range of the 7.62mm NATO is indeed 800m, as given in many military field manuals. This is roughly 880 yards. In comparison, the point target effective range of the 20"-barreled M16A2 is considered to be just over 600 yards.[19] For the M4 carbine the effective range is about 550 yards. This gives an advantage to the 7.62mm NATO in excess of 35 percent greater effective range. It is more often considered to be a 100 percent difference, meaning that in the days of the M193 5.56mm cartridge the maximum effective range was often given as 500 yards, compared to the often accepted 1,000-yard effective range of the 7.62mm NATO.[20]

This is a more realistic assessment of what these two cartridges are capable of. During the tests that led to the adoption of the M16A2 rifle and the M855 cartridge, the test results showed that it could outpenetrate the 7.62mm NATO M80 ball at 800 yards.[21] This was most likely due to the M855's steel penetrator tip. However, many question the validity of these results. Changing any test variables like target material and distance can yield vastly different results. While armor piercing ability may be more evenly matched between these two rounds, their effective range has proven vastly different, in terms of both hit potential and terminal effect.

As mentioned several times in this work, the current war in Afghanistan has shown severe shortcomings in the ability of the 5.56mm NATO to effectively engage the enemy beyond 300m.[22] This has long been known as a barrier distance for this caliber. At ranges beyond 300m, the 5.56mm round, in most cartridge configurations, rapidly begins to lose velocity and its trajectory suffers severely as a result. Inside of 300m, the 5.56mm is a fairly flat shooting cartridge, which makes for good hit potential given its velocity and relatively low recoil force, especially when fired from a soft recoiling weapon like the M16/M4. It also means troops don't need to mess with sight adjustments in the middle of a firefight. While no sight adjustment would be required for 7.62mm NATO rifle at these ranges, either, there will be a greater variance in point of impact at various ranges when holding center of mass as the aiming point. This of course depends on the initial combat zero of the rear sight. This variance is due more to the steeper trajectory of the much heavier 7.62mm bullet.

The traditional combat zero for the M16A1 rifle was 250m,[23] but it has recently become more popular to zero the current M4 series for 200m. This gives the least amount of shift in point of impact over the greatest portion of the 5.56mm's effective range when fired from this weapon.

The same is true of the 7.62mm NATO rifle when given a combat zero of 300 to 400m, with 400m likely a slightly better overall choice, though 200–300m is the more traditional setting. Three hundred meters was the zero range for the fixed rear sight on Dutch FAL variations. Within the limitations of iron sights, neither caliber should have much trouble hitting a man sized target from either service rifle. The problem arises at ranges beyond 500 yards and primarily from the 5.56mm NATO. The 7.62mm, at ranges beyond 500 yards, is still quite capable of accurate round placement. This becomes much easier with a properly zeroed scope. The same holds true for the 5.56mm, but it has lost so much velocity by this time that the slightest crosswind will throw it off its mark even if the elevation setting is correct. This is simply due to its far lighter bullet. Susceptibility to wind deflection was a big part of the switch from the original M193's 55 grain bullet to the M855's 62 grain bullet in 1982. The 62 grain bullet was intended to have far greater wind resistance than the original M193 round.[24] This is also why we are seeing a popular new, limited issue 77 grain bullet and its Mk 262 cartridge used in Afghanistan.

The Mk 262 is being used for reasons besides wind resistance, however. It carries its kinetic energy to a greater distance than the earlier two 5.56mm cartridges and offers superior terminal effect as a result. This is true even though it starts out at a much lower muzzle velocity than the M855. The Mk 262 has an initial velocity of 2,750 feet per second when fired from a 20" barrel of the M16 rifles series. In comparison, the M855 leaves the same barrel at over 3,000 feet per second.[25] The Mk 262 may be slower initially, but its heavier bullet is known for superior accuracy and engagement ranges are reported to be over 750 yards, bringing this round much closer, but not quite equal, to the 7.62mm NATO in performance, at least in terms of engagement range. However, it must be remembered that the 7.62mm NATO bullet would still have plenty of power in reserve at this range.

It must be considered that the performances reports concerning the Mk 262 are limited and, while most have been glowing, the same can be said of the recent M855A1 reports as well. The same grain of salt should be taken with both until the improved performance they claim to offer can be absolutely verified. It must also be remembered that there have been a few instances where the Mk 262 did not provide any noticeable increase in stopping power, such as the attack that took place in Ramadi, Iraq, in 2003.[26] It seems that the heavy 77 grain bullets did not help much during this attack, so the supportive reports must be considered carefully.

For many years, U.S. Marine Corps sniper teams have been required to make head shots at 800 yards during training with the Remington M40 series sniper rifles, with torso hits required to 1,000 yards.[27] Reports of 900-yard M4 shots in Iraq have been made[28] and this feat would not likely have been repeatable on any but the best of days. Targets don't usually wait around for the shooter to get it right.

The big concern with U.S. troops serving in the Middle East has been not only the shorter effective range of the 5.56mm but also its terminal effect when a target is hit. Reports have come back of combatants being hit multiple times by M855 bullets and yet continuing to fight.[29] By comparison, the lethality of the M193 round when fired from the original M16A1 was often reported to be good. This was especially true when fired from the early 1–14" twist barrels used during its early testing period in Vietnam. Again, the early reports that came back regarding the M16's first performance in Vietnam all but proclaimed it to be a super gun that blew targets apart, which should be considered a sales pitch more than anything else, although the pitch apparently worked on McNamara and the Kennedy administration.

This early rifling rate was of course later changed to the familiar 1–12" twist of the M16A1 due to the poor cold weather performance of the original 1–14" rifling. Even after the increased twist rate, the M193 cartridge performed fairly well, well enough in fact that some NATO member nations put off adopting the 5.56mm until a more humane bullet could be produced. There were complaints that the move to a 1–12" rifling resulted in a rather severe drop in wounding capacity.[30]

When the SS109 was standardized as a second NATO caliber in 1980, yet a third barrel twist rate was needed. The SS109 is Europe's equivalent of the U.S. M855. While some NATO members have gone to the 1–7" twist as found on the U.S. M16A2 rifle, others have opted for 1–9" barrel, which can effectively stabilize the SS109's 62 grain bullet yet provides better accuracy if using up old stock supplies of M193 ammunition.[31] This option was generally only a concern for those nations that had accepted the original M193 cartridge in the first place. The then new M16A2 was needed as the old 1–12" barrels of the M16A1 series just couldn't adequately spin the longer and heavier M855 bullet. Today the 1–9" barrel is proving more popular in the commercial market due primarily to their better accuracy with lighter bullets.[32] Some may ask about the heavier 77 grain bullet of the Mk262 and its stability when fired from a 1–7" barrel. Generally, as long as the barrel twist rate is faster than about 1–10 inches, stability is fine with overall accuracy of the heavier 77 grain bullet and generally better than accuracy with lighter bullet weights. It is not likely that the U.S. will ever adopt the current Mk262 Mod 1 cartridge as a standard round to replace the M855, despite its better long range performance. Its reason for not passing muster as a standard cartridge is of course its open tip match bullet design. While this design is not intended to cause expansion when hitting a target and is in no way an attempt to make the bullet a hollow point in terms of terminal effect, it doesn't meet U.S. standards in terms of barrier penetration and due to this shortcoming will not likely ever see adoption as a standard issue cartridge.

There is a new cartridge that has recently entered service and is, in fact, already the new standard issued to U.S. troops in Afghanistan. Reports are that the new M855A1 enhanced performance round (EPR) is more accurate than the M855 and has a higher muzzle velocity. However, it is also reported the chamber pressure has increased due to the use of a faster burning power to achieve this extra velocity. The EPR is also reportedly far superior in terms of its armor penetrating ability when compared to the M855. As the new bullet uses no lead in its construction, it is popularly being called "green ammo." One added bonus of this new cartridge is that it is also supposedly offers superior terminal effect against soft targets in addition to its better armor penetration performance.[33] All of this superior performance seems hard to believe, but time will tell if it can effectively do the job as claimed.

The new bullet weighs the same as the one used in the original M855, but construction is completely different. Gone is the lead core of the M855. Instead the new bullet uses a copper core, though the original test variation uses the core composed of a tin and bismuth alloy.[34] The tin was added to provide hardness, while the bismuth helped to add mass and sectional density. This alloy, however, proved to deliver inconsistent ballistics at higher temperatures. As a result, a new bullet design, using new construction material, was needed. The original M855 used an internal steel penetrator tip, while the new steel tip of the M855A1 is exposed and heavier than the original steel tip. Should the new M855A1 prove as effective as claimed, perhaps a heavier variation may prove an acceptable alternative to the Mk262 Mod 1 and its open tip match bullet. This would provide the U.S. with an extra long range 5.56mm round that could possibly come close to matching the 700m engagement ranges being delivered by the Mk262. However, it is my guess that some military members may have some other solutions in mind that take the form of some new and very unique calibers.

Recent conflicts have meant a change towards urbanized warfare and changing tactics dealing with this shift, the open expanse of Afghanistan being an exception. Possibly as a result of this change in warfare methods and tactics, there have been some new developments to recently appear in the ammunition industry in the form of new calibers designed for combat. However, the appearance of these new cartridges may have also been due to the complaints against the 5.56mm in general and the M855 in particular.

Some may ask that if the M16/M4 and the 5.56mm cartridge have served since the mid '60s, why the need to return to using the weapon and cartridge that were replaced by the more modern M16? The brief history lesson concerning the matter should have given everyone a fairly clear summary of how we came to the current situation regarding combat cartridge performance. For those who haven't quite made the connection, I will give a brief summary of the steps that have led to today's temporary solution of switching back to the 7.62mm NATO rifles currently being used in Afghanistan.

When the M16 first saw widespread use, it was during the close confines of jungle warfare. Its shortcomings for long range engagement were missed, as it was seldom used at such distances. At short ranges the 5.56mm and especially the M193 cartridge when fired from the early 1-14" to 1-12" barrels generally provides decent terminal effect, though not always. There were some complaints that even the M193 and its 1-12" barrel did not provide adequate stopping power. These stem from as early as the late '60s.[35]

The war in Vietnam came to an end and the matter never really surfaced for quite a while, at least not to the point where it would draw media attention, and negative media attention tends to be a primary motivator in a democratic society.

The following years ensured that the issue lay dormant for some time. The Cold War in Europe thankfully never heated up to the point where the M16A1 had to perform much beyond the rifle range at some U.S. Army base in Germany.

In the 1980s, there was Grenada and Panama. Both were decisive military operations of very short duration involving fairly short engagement ranges.[36] Even still, complaints about stopping power and the 5.56mm surfaced at times. These were followed by Operation Desert Storm, almost as short in duration as the operations of the 1980s. In retrospect, this operation was far less decisive than the other two.

There were more complaints about the performance of the 5.56mm, or rather its lack of performance.[37] By 1991 and Operation Desert Storm, the M855 cartridge and the M16A2 and its 1-7" barrel were in widespread use within the U.S. Army and Marine Corps. Thankfully again, Desert Storm was over quickly and the issue seemed to go away (to a lesser degree). Shooting started up again in Somalia for a short time in the early '90s, for those who remember. This involved fairly close range shooting against people who weren't all that bulky. The number of Somalis lost due to famine brought on by criminal behavior is too depressing to even bring up, so you get the point. Hitting a half-starved Somali rebel wasn't a representative example of the 5.56mm's capability with regards to terminal effect. Again, the problem was pushed aside after things cooled down.

However, after a shooting war that has lasted longer than any other military campaign in U.S. history, with the exception of the one against the Native Americans, the issue with the 5.56mm could no longer be ignored, and it hasn't been.

Solutions began to show up quite quickly once the shooting started in late 2001. By 2002–03, the Mk262 round was on the way.[38] The Mk318 SOST round made its appearance for the Marine Corps once the early M855A1 cartridge didn't live up to expectations.[39] These rounds have helped to a small extent and have alleviated some of the symptoms to a degree, but a cure has yet to be found in 5.56mm. These supplemental cartridges do not match the primary solution of reverting to using the 7.62mm NATO in place of the 5.56mm for current operations. That is not to say that the cure does not come without its drawbacks. The 7.62mm is far heavier and the troops are not able to carry the same combat load of ammunition while on patrol. Also, the recoil of the 7.62mm NATO is far more severe and this has a rather severe effect on those who are recoil sensitive. This leads to training issues as well. Though for the special operations units this is less of a problem, as these men tend to train with rifles far more regularly and quickly learn to deal with the recoil issues.

It is not even a matter of debate that there is only so much the 5.56mm can be expected to do, especially at ranges in excess of 300m. Anyone who tries to insist that everything is fine is only kidding themselves. This is not a good idea where the lives of U.S. service members are at stake. While it is true that the majority of combat situations likely to be encountered are urban in nature, the war in Afghanistan is the rare exception.

As many already know and as covered in other places within this work, the combatants in Afghanistan learned quite quickly to exploit the advantages of their terrain and weapons to best serve them. They observed early on the shortcomings and limitations of the 5.56mm and have adapted their engagement practices and tactics to take full advantage of the situation. This is only to be expected, and any enemy would be foolish not to do so.

The primary weapons in use by the Taliban in Afghanistan are Soviet in origin. These include the AK-47, primarily in the most common AKM variant with its stamped steel receiver. This would include the folding stock AKMS model. Both of these are chambered for the 7.62×39mm Soviet cartridge (M43). There are also a fair number of 5.45×39mm AK-74 variations in use. These are presents left over from the 1979 Soviet invasion and consequent occupation of the country. Other leftovers include the fairly common and very lethal SVD Dragunov 7.62×54mmR sniper rifle. This rifle is one of the most common sniper weapons in existence and can provide accurate and effective fire out to over 1300m (1,422 yards).[40]

The primary general purpose machine gun (GPMG) in use by the Taliban is the 7.62×54mmR PK (Pulemyot Kalashnikova). This usually takes the form of the lightweight infantry model known as the PKM. This is a modernized PK variation using a stamped steel receiver to help save weight. The end result being a belt-fed (right side), bipod equipped machine gun with an empty weight of around nineteen pounds. This makes it considerably lighter than the FN MAG or U.S. M60E1, which have been its traditional rivals over the years.

The PKM is often considered to be more reliable than the M60 and as reliable as the MAG.[41] Its only real disadvantage is that it utilizes a non-disintegrating link belt for feeding. From a tactical perspective, this can be a bit of an impediment. A long, dangling spent belt can be a snag hazard, especially in heavy brush. This, however, can be quickly resolved with a pair of metal cutters.

The PKM fires an extremely old cartridge of a design that dates back to 1891.[42] This is an old-fashioned cartridge that uses a rimmed case. This is one of the few cartridges that managed to survive the rimless cartridge changeover that began in earnest at around the end of the nineteenth century.

The .303 British, along with the SMLE (Short Magazine Lee Enfield) rifle that fired it, also weathered through and wound up serving for another fifty-plus years, some of those in Afghanistan, against the Soviets no less. The fact remains that while the 7.62×54mmR is a very old cartridge with a fully rimmed profile, it is actually slightly superior to the 7.62mm NATO where ballistics are concerned. Velocity is slightly better and bullet weights are comparable. The 7.62×54mmR is capable of providing lethal area fire to a range of 2000m.[43] Due to the rimmed case design, both magazine and belt-fed designs had to be adapted to provide reliable feeding. This presented a bit of a challenge for both the SVD and the PK series.

While the PKM and SVD both provide the enemy long range engagement capability, the standard AK-47 7.62×39mm cartridge is no slouch. While the traditionally accepted effective range for the 7.62×39mm is 400m (440 yards), the bullet remains lethal at far greater distances.[44] It must be remembered that while it only starts out at a little over 2400fps, the AK's bullet weighs roughly twice that of the M16/M4. In comparison, the 5.56mm M855 starts off at around 2,800 fps from an M4 barrel and the bullet is only 62 grains, compared to the 123 grain bullet of the M43 cartridge most often used in the AK-47/AKM series. This gives the 7.62×39mm a great advantage in terms of retained energy and momentum.

This selection of small arms, combined with the 82mm mortars and RPG-7s used by the Taliban, allows the enemy to begin their attacks at ranges of over 4,000 yards (82mm mortar).[45]

The RPGs used can send warheads flying to ranges of over 1,000 yards

This is the U.S. M110 Semi-Automatic Sniper System. This 7.62mm NATO rifle is really just a very well-made modernized AR-10. The AR-10 was the parent rifle of the current U.S. M16 series, which includes the M4 carbine. As both weapons use similar gas systems, they suffer from the same shortcomings, though the semi-automatic rifles tend to get dirty far more slowly. (Spc. George Hunt)

in some cases, and as mentioned, the PKM can start peppering U.S. troop locations at ranges to 2,000 yards. Once the ranges close to less than 1,000 yards, small arms fire starts getting very dangerous. This is within the point target capability of the larger .30 rifles like the SVD and our own M14s, M110s, and SCAR-Hs. There have been reports that the M110 semi-automatic sniper rifles have some reliability issues, however.[46] As these are basically bigger versions of the M16, this is not really all that surprising. It is unlikely that the reliability of the M110 can match that offered by the SCAR-H or M14, simply by virtue of its design. Reports that some U.S. troops have been substituting the M110 for the SCAR-H, even though it is less accurate, are testament to reliability issues with the direct impingement gas system. The Taliban also likely has a few remaining .303 bolt actions left over from earlier wars. These old SMLE rifles have an effective range similar to that of both the 7.62×54mmR and the newer 7.62mm NATO.

With engagement ranges of less than 800 yards, the AK rifles can start hitting U.S. troops, though not with any great consistency, though U.S. troops might want to argue this matter.

The Soviet heavy caliber 12.7mm and 14.5mm machine guns are not mobile enough for the Taliban to easily move in hillside operations that often involve setting up hasty ambushes. These heavy machine guns can be horribly effective as defensive weapons and can provide lethal area fire in excess of 4000m (especially the KPV 14.5mm).[47]

In comparison, the U.S. troops operating in Afghanistan are primarily equipped with M4 carbines and M16A4 rifles, M249 SAWs, and M240 series machine guns. This is in addition to the larger 7.62mm rifle mentioned earlier. Heavier firepower for U.S. troops comes in the form of the SMAW (Shoulder Launched Multipurpose Assault Weapon) and the M224/

M224A1 60mm mortar. The M224 is often used without its bipod, as this is a far quicker deployment method, like when returning fire during an ambush. There is reportedly a lightweight version of the U.S. 81mm mortar in development that will provide U.S. troops with the capability to match the larger 82mm Soviet/Chinese/Pakistani mortars being used by the Taliban.[48] These larger mortars offer greater range and larger explosive effect.

Given the terrain in Afghanistan, this is about the extent of what troops can carry on most operations. It is under these conditions that the M4 carbine falls short of its task as the primary U.S. Army shoulder arm. The M16A4 used by the Marine Corps doesn't fare a whole lot better, for that matter. The new M855A1 is reportedly helping to correct the terminal effect issues of the M855. Again, this has yet to be proven to a degree where it can be considered gospel. Government reports of the past have been known to embellish just a bit. Should the M855A1 prove to be less than promised, the Mk262 does provide troops with a longer engagement capability, approaching 700m by most accounts. Given this considerable extension to the M4's effective range, it is likely that this round is much appreciated by U.S. units that are fortunate enough to be issued this cartridge.

One bright spot to all of this is the generally better accuracy of the M4 than that of the AK-47/AKM. With shooting ranges being extended out past 300m, the better accuracy helps improve the chances of U.S. troops in hitting Taliban members. The usually poorer accuracy of the AK series puts it at a slight disadvantage here. This hopefully gives U.S. units a somewhat better chance in gaining the initiative when attacked from higher elevations, which is usually the case, as the U.S. troops are usually forced to pursue.

The presence of a greater percentage of M14s or SCAR-H rifles improves U.S. capability under these combat conditions to no end. The only real limitation at that point becomes the shooter's grasp of the 7.62mm NATO's ballistic behavior at extended ranges where thorough knowledge of trajectory and accurate range estimation are an absolute must if accurate shot placement is expected.

With improvements in 7.62mm NATO bullet design such as that of the M80A1 updated cartridge, performance should only improve, though it already holds a definite advantage over the 5.56mm in terms of effective range, past military test results aside. Any of those who remained in denial over all these years, who desperately wanted to believe the many military test results conclusively "proving" the 5.56mm's superiority over the 7.62mm, have pretty much been silenced by the issues that have surfaced in Afghanistan during the last eleven-plus years of fighting. When the U.S. Army's professionals start submitting formal papers on the matter (like Major T.P.

Ehrhart), the issue is serious enough to take a very close look at. This is precisely what prompted development of the 6.8mm SPC, among other possible solutions.

The first new caliber up is the 6.8mm Remington Special Purpose Cartridge (SPC). This cartridge is also sometimes known as the 6.8×43mm. The development was a joint venture between Remington Arms Company, Inc., and the U.S. military.[49] The U.S. military wanted a more effective round than the 5.56mm in terms of stopping power and barrier penetration. They also wanted the round to be able to fit the dimensions of the M16/M4 series magazine. This meant limitations in the overall cartridge length. The Remington team started with the old .30 Remington case, which was then shortened and necked to accept the bullet of the 6.8mm. This cartridge offers performance superior to that of the 7.62×39mm Soviet AK-47 cartridge. Bullet weights are similar to that of the 7.62×39mm, but the bullet of the 6.8mm offers better performance in terms of velocity retention due to a higher ballistic coefficient in addition to its higher initial muzzle velocity. The 6.8mm generally leave the barrel around 200 feet per second faster than the 7.62×39mm bullet. The 6.8mm offers far greater hitting power than the 5.56mm and yet operates at a lower chamber pressure. In addition, the 6.8mm is available in several standard bullet weights ranging from 85 grains up to 140 grains, with 110 to 120 grain bullet weights being the most often seen in use for combat purposes. The 6.8mm was designed to perform optimally from a 16" barrel, with no excessive muzzle blast and flash like that which occurs when the 5.56mm is fired from the shorter barrel of the M4 carbine.[50]

The lighter bullets used in the 6.8mm deliver even higher velocities than 120 grain variations. The 110 grain bullet weight is one of the most efficient. This bullet is moving at roughly the same velocity at 400 yards as is the 77 grain bullet of the Mk 262. It must be remembered, however, that this bullet carries with it a much larger diameter and considerably more weight than any 5.56mm projectile. Bullet drop at these longer ranges is also similar for both cartridges. One disadvantage to the new 6.8mm cartridge is that it requires some considerable changes to convert the standard 5.56mm M16 or M4 in order to take advantage of the hard hitting 6.8mm bullet. Technically, only the bolt, barrel, and magazine are different. For most caliber conversions, it is easier to replace the entire upper receiver with a new assembly and then replace the magazine. This can be done in about a minute. The standard thirty-round M16 magazine, however, will not hold the same number of rounds when manufactured for the 6.8mm. Only twenty-five to twenty-eight rounds of 6.8mm can be loaded into a magazine of these dimensions, depending on the manufacturer.[51] While simply swapping out the upper receiver is a pretty typical conversion for the M16 pattern weapons, there is another new caliber we will be discussing that simply requires a barrel change. While this new caliber doesn't offer quite the same power as the 6.8mm, it does bring with it some other advantages.

The 6.8mm SPC is a very likely candidate for a possible 5.56mm replacement in the future. Before anything like this occurs, a great many things will have to happen. First, the U.S. economy will need to improve a great deal. This is something that hadn't really happened as we headed into the fifth year of the worst economic position since 1932. Second, U.S. involvement in the Middle East will have to be pulled back considerably. With the current instability and tension in this region, this may not happen as quickly as we would all like. Finally, a great deal more testing will have to be done to truly determine the best possible choice for a 5.56mm replacement, as there are several good candidates currently available, and future development may lead to an even more effective cartridge design.

This last factor is easier said than done. Given the amount of money invested in 5.56mm ammunition production and the cost of developing an entirely new family of weapons, we are not likely to see the new cartridge adopted anytime soon. Also, while other NATO member nations have lately developed the same opinions regarding the 5.56mm's performance, many of these nations are in no better financial position than we are to undertake such a caliber move. This is largely why we've seen a move back to the 7.62mm NATO round for use in Afghanistan. This was NATO's first standard caliber after all, and many nations still have large inventories of both weapons and ammunition in this caliber. However, the desire to move to an intermediate cartridge/weapon combination that offers less weight and recoil than that of the 7.62mm NATO weapons family and yet offers far greater range and lethality than the 5.56mm NATO is strong among some military people both here in the U.S. as well as in other nations. Since those who would like to see this move are usually those who use the weapons the most in combat, their opinions matter greatly and we may need to listen to their voices if we want our sons and daughters coming home safe and sound and in one piece.

The next cartridge to be covered is potentially an even stronger candidate as a possible replacement for the 5.56mm NATO, the 6.5mm Grendel, also known as the 6.5×38mm. This 6.5mm has roughly the same overall dimensions as the 6.8mm. However, it possesses a much greater long range potential. Both cartridges were introduced around 2004 and offer similar bullet weight options and muzzle velocities. The 6.5mm Grendel has a reputation for not only accuracy but velocity retention at longer ranges as well. Like the 6.8mm, this round is capable of being fired from any AR (M16) type rifle and magazine well. Magazine capacity with this cartridge, however, is reduced to twenty-six rounds, even fewer than that offered by the 6.8mm.[52] To many, this reduction in magazine capacity is less important than the increase in performance offered by the new cartridge. Changes required to convert an M16 to fire this new cartridge are the same as for the 6.8 mm conversion, meaning a new bolt, barrel, magazine, et cetera.

The 6.5mm Grendel is not a new cartridge, with the design having been finalized by 2003. This was a joint project involving competitive marksman Arne Brennan, Bill Alexander of Alexander Arms, and Janne Pohjoispaa, an engineer from the Finnish ammunition manufacturer Lapua. This cartridge was not standardized by the Sporting Arms and Ammunition Manufacturers' Institute (SAAMI) until 2012.[53]

The 6.5mm starts out moving at roughly the same speed as the 6.8mm, and this combined with superior bullet geometry (high ballistic coefficient) leaves the 6.5mm bullet moving along at supersonic speeds for a great distance. The cartridge case of the 6.5mm Grendel has a much sharper shoulder angle than that found on the 6.8mm SPC, and this has generally been associated with the better accuracy of the 6.5mm. Accuracy is something that the 6.5mm Grendel has in spades. At the same time, a sharp shoulder angle often presents extraction issues, especially in self-loading weapons. Thus far, this has not been a major issue with the 6.5mm; however, more thorough testing will have to be done to determine this issue conclusively. One of the most unique aspects of the 6.5mm cartridge is its efficiency. While it is an intermediate cartridge, it has a ballistic coefficient (BC) in excess of 0.600 for some projectiles in this caliber.[54] While many readers of this book may be familiar with this term, a quick explanation may be in order. A ballistic coefficient is a number value assigned to every projectile designed; the higher the number, the better the bullet retains its velocity and resists wind deflection. According to most tests, the 6.5mm can retain its velocity like few other rounds. It holds its velocity so well, in fact, that it retains greater kinetic energy at 1,000 yards than the M80 ball round of the 7.62mm NATO cartridge.[55] It does this with a lighter, shorter cartridge, which possesses far less recoil and which is capable of being fired from an M16 platform.

The 6.5mm Grendel is reportedly a versatile round, and some AK pattern rifles chambered for the 6.5mm are beginning to hit the commercial market. In a sense, the round is merely returning home, as this cartridge was initially based on the .220 Russian cartridge, which was designed in the 1950s for target shooting. In turn, the .220 Russian cartridge was derived from the 7.62×39mm Soviet AK-47 cartridge.[56]

While the 6.5mm Grendel and the 6.8mm Remington SPC are the two most likely candidates for a replacement intermediate cartridge, other new cartridges recently have appeared. One other new cartridge may have been designed in direct response to yet another Russian cartridge that has been proving quite effective in dealing with urban combat conditions in recent years.

The Russian 9×39mm is a specialized subsonic round that fires an extremely heavy 260 grain bullet at roughly 1,000 feet per second. This cartridge allowed for the development of a family of short, lightweight rifles with which to fire it. The end result is basically a submachine gun size package in terms of weight and length, with some being shorter than fifteen inches with their stocks folded. Empty weights for some of these weapons are less than four and one-half pounds. Yet while this cartridge is used in weapons of such diminutive size, the power being delivered by the 9×39mm is anything but submachine gun-like. While the 9×39mm muzzle velocity and bullet weight is not all that different from that of the .45 ACP, its velocity retention is far greater. This is because the cartridge uses a spitzer type rifle bullet resulting in downrange performance that is quite different from that of a pistol cartridge like the .45 ACP, when fired from a submachine gun. In fact, the 9×39mm possesses roughly twice the momentum of the 9mm Parabellum when fired from weapons of similar size.

Since first introduced in the mid-'90s, this cartridge, when combined with the various weapons that utilize it, has given Russians an excellent all around urban combat weapons platform. The original family of weapons was the AS, VSS, and SR3, which has sometimes been called the MA. The AS was designed as a suppressed assault rifle and should never be fired without the suppressor installed, as doing so can result in severe injury or death of the user. The same goes for the VSS silenced sniper rifle. The VSS is basically the same weapon as the AS but uses different furniture. The stock and scope of the VSS are quickly detached and the entire weapon is meant to be carried disassembled in a special case designed for this purpose. The final model mentioned, the SR3, was designed as a non-suppressed compact assault rifle. The SR3 uses a very compact top folding stock, as opposed to the side folding tubular metal stock of the AS. There is also a newer version of this rifle known as the SR3M, which uses a folding stock similar to that of the AS. The newer version can also accept a detachable suppressor.[57]

There is another compact top folder available that is very similar in size and weight to the SR3. The 9A91 is apparently more commonly seen than the SR3, and this is likely due to its lower production cost. In addition to the 9A91, there is a VSK94 variation designed as a sort of paratrooper version of the VSS. Unlike the VSS, the VSK94 can be safely fired without its suppressor installed.[58]

There is also the new AK variant chambered for this round that was introduced a few years back and is known as the AK-9. This model is very similar to other weapons in the AK-100 series carbine family. The AK-9 was intended to be used with a detachable suppressor but may also be safely fired without the suppressor installed.[59]

This family of weapons and their hard hitting cartridge have given the Russians a perfectly suited weapons platform for the job of urban combat, and watching this development must have made some within the U.S. intelligence community rather nervous.

Perhaps this is unrelated, but a new cartridge has recently appeared in a big way in the U.S. commercial market. This new cartridge is suspiciously similar in performance to the 9×39mm Russian round, though this new cartridge does offer some rather big advantages, which we will discuss shortly. The new Advanced Armament Corporation (AAC) .300 Blackout cartridge has become popular and has done so rather quickly. This cartridge is also known as the 7.62×35mm or more often as the .300 BLK.[60] It is perhaps unfair to say that the reason for the .300 BLK's newfound popularity is a direct result of the need to counter the Russian 9×39mm, as the new .300 BLK cartridge is quite likely a superior overall round, though only time will tell, as it has only recently seen any service.

The .300 BLK is technically not a new cartridge. Its development was largely influenced by the .300 Whisper round developed in the early 1990s by SSK Industries. In fact, there's very little difference between the two cartridges. The .300 Whisper cartridge has been around for some years and has become one of the more popular rounds in the Whisper family lineup. The Whisper cartridges were designed as subsonic rounds that offered heavy and efficient bullet weights largely for suppressed shooting. These have been developed in calibers up to .50.[61]

Once standardized, the .300 BLK was offered in factory loadings with performance quite similar to that of the slightly larger Russian 9×39mm. The .300 BLK uses a standard subsonic loading with a 220 grain bullet, as opposed to the 260 grain slug of the 9×39mm. In addition, the smaller diameter of the .300 BLK will likely offer superior velocity and energy retention downrange. This really depends on the ballistic coefficient values of the various bullets that may be used in either cartridge.

There is one big difference, however, between these two "urban combat" rounds. The .300 BLK is available as a supersonic intermediate cartridge loading in addition to its more easily suppressed subsonic load. This .300

BLK intermediate load is quite lethal, though it does not quite meet the performance level of the 6.8mm SPC or 6.5mm Grendel. The .300 BLK supersonic loading is more along the lines of a 110 to 125 grain bullet moving at a muzzle velocity of between 2,100 and 2,400 feet per second, depending on barrel length and the particular bullet weight being used. This type of performance puts it quite close to the 7.62×39mm Soviet AK cartridge.

While the .300 BLK does not perform quite as well as the 6.8 or 6.5mm, it offers yet another huge advantage over either of these cartridges, its external dimensions. As with the 6.8 and 6.5mm, the .300 BLK can be used in any AR pattern weapon and magazine well. However, only with a .300 BLK can the actual 5.56mm M16 magazine be used, along with the same bolt and return spring and buffer. In fact, the only part of any M16 pattern weapon requiring change for conversion to .300 BLK is the barrel itself.[62] Given this characteristic, it's a shame that the M16 family doesn't offer a quick change barrel feature. If the M16 came so equipped, the .300 BLK would probably already have been adopted as an additional standard cartridge for U.S. troops.

The .300 BLK can provide a quiet, lethal, heavy subsonic slug capable of hitting accurately out to around 300m when used in combination with an effective suppressor. At the same time this cartridge can provide an effective intermediate loading, which can lay down accurate and lethal fire out to 500 yards. This is not far behind that offered by the 6.8mm SPC but not quite as good as the long range performance of the 6.5mm Grendel. The versatility of the .300 BLK is unmatched by either of the other calibers, however. While the three new cartridges discussed in this chapter likely represent the pinnacle of possible contenders for replacing the 5.56mm, this is not likely something to happen anytime soon. There are far too many financial burdens in making such a costly move as changing to a new standard cartridge. This is especially true with cartridges that are so new and as of yet relatively untested. As of now, these cartridges have seen minimal combat use, but the 6.8 SPC has reportedly been tested in Afghanistan with favorable results.[63] Should these new rounds see some testing within the military, we may develop a more complete picture of what they bring to the table as a whole. As it stands, the 6.8 SPC is generally agreed to be the most effective round at ranges of less than 400m, with the 6.5mm being a better round, by a good margin, at ranges beyond 400m. The maximum effective range of the .300 BLK in its supersonic loading is just beyond this distance, but the .300 BLK was clearly designed for much closer work and offers unmatched versatility and flexibility that allows it to meet often changing tactical needs.

There is one concern regarding the 6.5mm Grendel in that it doesn't appear to offer the same short barrel performance as the other two rounds.

The 6.5mm Grendel appears to perform best with barrels in the 20–22" range. This is something to consider thoroughly, as it is likely that carbine barrel lengths will be used as standard for future issue shoulder arms in the U.S. military. This is due to the carbine's ease of handling in most urban conditions.

From a practical point of view, the primary concern from the start, involving the development of these cartridges, was the terminal effectiveness, or rather lack of terminal effectiveness, of the 5.56mm NATO cartridge, along with its poor long range engagement capability. The 6.8mm SPC offers improvement in both categories, while the 6.5mm Grendel offers similar improvements in terminal effect but is a vastly better long range cartridge than the 6.8mm. From this point of view alone, the 6.5mm would seem to be the better choice. One factor that gives the .300 BLK a big advantage or the other two cartridges regarding possible consideration for adoption is the lower cost involved in converting the M16 platform in order to fire the new cartridge.

It is a reasonably educated guess that we will see the .300 BLK used by our military on a limited basis, with possible continued work on improving the 5.56mm cartridge in terms of both its terminal effectiveness and its long range engagement ability. This is simply due to the exorbitant cost of switching to a new standard caliber like the 6.8mm SPC or 6.5mm Grendel. It is likely that the 5.56mm has come about as far as it can with the new M855A1 and Mk 262 Mod 1 cartridges.

Since all three of these cartridges were designed to be used in the M16 platform, this is in no way a limiting factor regarding options for selecting an adequate replacement cartridge for the 5.56mm. If the U.S. is going to spend the kind of money involved in switching over to a new standard caliber, it would seem prudent to start from the ground up and look at other possible options regarding the weapon platform itself. It isn't written in stone that the M16 system be kept as the U.S. standard rifle. The overall cost of changing to a new rifle design, in truth, isn't going to be that much greater than changing the upper receiver and magazine of the M16 platform. The cost of a complete lower receiver assembly with stock is roughly 25–35 percent of the total cost of an AR pattern weapon.[64] If the U.S. is to undertake such a project, it should use that opportunity to get the results wanted and needed by our troops, a truly reliable rifle system using a highly effective intermediate, low recoil cartridge that possesses both good terminal effects and long range engagement ability. Perhaps the next system will prove more consistently reliable in stopping combatants at longer ranges and, at the same time, providing an increased level of confidence for U.S. troops.

An educated guess is that the least costly approach right now would be the design and standardization of a more effective 5.56mm cartridge for general issue, with the long range 7.62mm NATO as a limited standard for extended fighting conditions like the ones our troops are facing in Afghanistan. These conditions would apply to warfare in environments such as desert, arctic, mountainous terrain or plains, basically anyplace where open terrain is dominant and shooting is likely to take place at extended ranges beyond 500 yards. Some argue against keeping the 7.62mm NATO cartridge due to its recoil and lack of control when fired on full-auto. This is not really a valid argument, as firing any lightweight weapon in full-automatic mode, no matter what caliber, will be marginally effective at long ranges. This type of fire is primarily effective for suppressive purposes and is extremely wasteful of ammunition. Since many operations in this type of terrain require travel on foot, troops are required to carry this ammo in and are limited in how much they can carry. However, when fired in semi-automatic mode the 7.62mm NATO cartridge has historically proven highly effective and the M14 currently being used is well known as being one of the most accurate service rifles ever developed. Bullets aren't much good unless they hit what they are aimed at. As it is already in the supply chain, the 7.62mm NATO will likely remain in use for some time to come.

There is a definite tactical place for the .300 BLK. As this cartridge requires no modification other than a barrel change, it is our lowest cost option for caliber conversion from the 5.56mm to one that offers excellent barrier penetration along with effective suppressed fire for urban operations and CQB. Given its versatility, it is highly likely that this cartridge will see a good deal of use in the future regarding select U.S. military operations and will likely have a place within the law enforcement community as well.

The last subject of discussion in this chapter deals with the actual performance of these rounds. While the 7.62mm NATO was clearly an effective long range round, this does come with a severe price, its strong recoil force. The 6.5mm Grendel and 6.8mm SPC both generate roughly half the recoil force of the 7.62mm NATO. The .300 BLK generates recoil forces only slightly greater than that of the 5.56mm.[65] What this means to the user is quicker recovery time for a follow-up shot. The 7.62mm NATO also suffers a disadvantage in that both rifle and ammunition in this caliber generally weigh more, though here we are mainly comparing it to the 5.56mm NATO cartridge and its accompanying rifles, so this is arguable, as there is a great variance in weight among the many 5.56mm rifles available. The weight advantages offered by the larger intermediate calibers like the 6.8mm SPC, 6.5mm Grendel, and even the .300 BLK, are less defined here due to the significant increase in their bullet weight, though the powder charges of these rounds is less than the 7.62mm NATO, as is the weight of the cartridge case itself.

While the 7.62mm NATO cartridge can effectively engage targets to 1,000 yards, any rifle being used with iron sights will likely only prove useful to roughly half this distance. This is of course in terms of the average user's ability to engage individual targets. Things have changed considerably in this regard, with the insanely rapid rise in the common military use of optical sights. Compact, rugged, and sometimes relatively inexpensive, magnified optics seem to be universal in their use today. Many of the new optics include built in ranging reticles combined with bullet drop compensation. This combination easily allows the user to take full advantage of the maximum effective range of a cartridge. This is part of why we are seeing a resurgent use of the M14 in Afghanistan. This rifle is accurate to begin with and still packs considerable power at 1,000 yards. When it is used in combination with an effective tactical scope like the Trijicon ACOG or any of the other large number of ranging scopes available, the effects are lethal. These scopes are not only far more durable than the optics of the previous generation, but they are also far quicker to use, depending on the reticle included. Auto ranging reticles allow for engagement at various ranges without any major adjustment or guesswork involving aiming holdover. The Shepherd scope is one of the best examples of a modern optic using a built in ranging reticle. This scope is zeroed for 100 yards and has separate aiming points built in to 1,000 yards in 100-yard increments.[66] The user merely ranges the target by placing it in the proper fitting circle, holding the sight picture and firing. It doesn't get any faster than that.

The standard U.S. M118LR round drops to a subsonic velocity at roughly 1,000–1,100 yards. By comparison the 5.56mm Mk262 cartridge drops to a subsonic velocity at roughly 800–900 yards. When these bullets hit the transonic velocity range, they begin to suffer severe instability and accuracy is drastically affected. By comparison, the 6.5mm Grendel can maintain a supersonic velocity out to nearly 1,300 yards with its heavier bullets in the 120–140 grain range.[67] This is due to the far superior aerodynamic properties of the 6.5mm bullets, and the 6.5 mm is able to do this even though the bullets are starting out slower than either the 7.62mm NATO or the 5.56mm. The big advantage here is that the 6.5mm is able to do this with less recoil than the 7.62mm NATO, which by default results in faster and more accurate follow-up shots, especially at long ranges.

The 6.8mm SPC and .300 BLK cartridges cannot match the 6.5mm Grendel in this regard, but they do offer enhanced performance over the

5.56mm in other important areas. The subsonic capability of the .300 BLK aside, both rounds offer considerably better performance with regard to terminal effects than that provided by the 5.56mm. In addition, barrier penetration is better with both rounds due to their far heavier bullet weights. This last issue has been a major complaint regarding the 5.56mm and its difficulty punching through the windshields of cars being piloted by suicide bombers. This was the primary motivation for the development of the hard hitting yet short range .50 Beowulf.[68] Again, some of the issues regarding the terminal effects of the 5.56mm round can be addressed with a well-designed bullet in the 70–80 grain range, similar to the weight of the bullet used in the Mk262. However, the terminal effect would have to be top notch.

With the 5.56mm, its velocity is so high that the 62 grain bullet of the M855 tends to fragment on impact at close ranges when contacting hard objects like glass. In fact, it tends to yaw and fragment at close range on soft tissue as well. This tendency to fragment is why our troops have had issues regarding its ability to penetrate barriers in general, but it is also credited with giving the round its stopping ability at close range as well. It was also the motivation for the USMC's use of the Mk318 cartridge and the M855A1's adoption as our new standard issue the 5.56mm round.[69]

The original M855 bullet tends to hold together and penetrate targets most efficiently at roughly 200 yards.[70] By this time, its bullet has lost enough velocity that the fragmentation effect has dissipated and the bullet holds together more readily. This is actually a characteristic of many traditional high-velocity bullets. The M855A1 is reportedly far better in terms of hard target penetration and maintains its terminal effectiveness at long ranges as well as close.

While the 77 grain bullet of the Mk262 gives up some ground in regard to its ability to penetrate barriers, its primary reason for being developed was not to punch through car windshields but to allow for engagement beyond the effective range of the M855 round, which is limited to around 600 yards against point targets, and that's on a good day.

As for the potential 5.56mm competitor, the 6.8mm SPC, its ability to penetrate barriers is more reliant upon the weight of its bullet, but performance would no doubt be enhanced with the use of a tungsten cored armor piercing bullet in this caliber, which has already been developed. The same goes for the 6.5mm Grendel. The 6.8mm bullet is apparently capable of roughly the same level of armor penetration as the M993 7.62mm AP round.

The limitations in the performance of the 5.56mm are only natural, as it is unrealistic to expect what is basically a .22 caliber bullet moving at 3,000 feet per second to equal the performance of a larger diameter bullet like the

.30 caliber, which weighs two to three times as much (depending on comparative loads) and moves at a velocity only 200 feet per second slower.

In comparison, the 6.5mm Grendel is pushing its 123 grain bullet at 2,400 feet per second (from a 14½" barrel). At first glance, this seems like far inferior performance when compared to the 6.8mm, which pushes a 115 grain bullet to 2,600 feet per second muzzle velocity, from a similar barrel length. However, the same loads at 400 yards have dropped to roughly 1,800 feet per second for the 6.8mm, while the 6.5mm begins to edge out the 6.8mm in performance, having only dropped to 1,830 feet per second.[71] From this distance on, the 6.5mm becomes the superior projectile due to its much slower velocity loss. The differences become more pronounced as the range approaches 1,000 yards. With changes in propellant, it is quite likely that better short barrel performance can be squeezed out of the 6.5mm. Any change in propellant, however, has a tendency to lead to a consequential change in chamber pressure, sometimes leading to unsafe levels. Another concern is that an increase in chamber pressure may also adversely affect the inherent accuracy of the 6.5mm, which is a crucial component to its overall performance as a high-performance long range intermediate cartridge. All of this of course is merely academic, as we are likely years away from a possible standard caliber change.

For now U.S. troops are primarily using the new M855A1, which is reportedly performing quite well in the field. This is of course based on a limited number of reports coming back from Afghanistan, though there are also arguments that this new round is not nearly as effective as is claimed and, in fact, may lead to highly accelerated wear on the M16/M4 series. According to the U.S. Army, this new 5.56mm cartridge provides far better hard target penetration. This new round is reportedly capable of penetrating three-eighths-inch-thick mild steel plate at a range of 350m. By comparison, the original M855 could only achieve this level of penetration at roughly 160m. While both cartridges outperform the 7.62mm NATO M80 ball round in this regard, it must be remembered that the bullet of the M80 is constructed of lead and jacket material only, with no steel penetrator present. Performance for either the M993 7.62mm tungsten core round or the new M80A1 (same bullet design as M855A1) is no doubt far superior with regards to armor penetration. Just for comparison, the 5.56mm M995 tungsten core round is reportedly able to penetrate 12mm of armor plate at 100m, while the 7.62mm M993 can punch through 18mm of the same armor at the same distance.[72]

The M855A1 may offer improved performance over the M855, especially in regard to its armor penetration ability. This increase in performance does not come without a price. The extra cost per round will quickly add up, as the new round is reportedly far more expensive. An even bigger problem is likely to arise from the exposed steel penetrator tip of the new cartridge. The feeding of these cartridges into the M16 may lead to accelerated wear in the feed ramp area, as the lower receiver is constructed from aluminum and part of the feed ramps is cut into the receiver. It is possible, however, that the bronze coating over the steel penetrator tip may slow the rate of this wear a bit.[73] The bronze coating is installed to prevent corrosion of the steel penetrator, which would otherwise quickly occur after handling. In any case, the improved armor penetration and velocity are definite improvements, along with reportedly superior accuracy. The primary disadvantage of the M855A1 is the increase in chamber pressure (nearing maximum allowed for the M16/M4), due to the switch to a faster burning powder, which was done to increase muzzle velocity when fired from the shorter barrel of the M4 carbine.[74] As for the added cost, the cartridge may prove worth it if the enhanced performance helps increase the safety of our troops, though this has yet to be proven. There have been some concerns regarding possible decreased barrel life due to the use of the new round, but definite conclusions on this matter do not appear to be well proven yet. One concern may be the new bullet's construction method, as it seems clear the bullet was designed to separate upon impact. The strong argument can be made here that if the bullet was designed to fragment it may very well violate the Hague accords. Once more, proving this over the years has been impossible, as evidenced by the U.S. M193 cartridge and the Soviet 5.45×39mm original ball cartridge (pattern 74).

As a final note on current issue ammunition and potential replacements, the new 7.62mm NATO M80A1 will most likely replace the long serving M80 ball cartridge. This decision will most likely depend on whether or not the new cartridge drastically accelerates barrel wear in the remaining M14s, along with other U.S. 7.62mm NATO rifles. The increased cost of the new 7.62mm round will also be a factor.

The M855A1 will most likely remain the U.S. standard issue cartridge in 5.56mm for some time. This is of course unless it proves too harsh on the weapons due to pressure levels and possible barrel erosion issues. The ideal 5.56mm cartridge would possess the long range engagement capability of the Mk 262 (700m) and perform as well or better than the M855A1 in terms of its terminal effect, accuracy, and barrier penetration, piece of cake.

The 7.62mm will continue to see limited use, possibly more so now that an increasing number of individuals within the defense community have come to face the reality that the 5.56mm M855 (and possibly M855A1) never really lived up to its hype and there is only so much that can be done to improve the cartridge's performance. It is not entirely certain that the M855A1 represents the pinnacle of what could have been accomplished given our potential for product improvement.

As for a possible replacement intermediate cartridge and new standard caliber, the 6.5mm Grendel does indeed appear to hold the greatest overall potential (especially as a long range performer with light recoil); however, there is still a great deal of research remaining to be done to fully determine this. This is especially true regarding its shortcomings in carbine length barrels. There is also some concern regarding its flexibility in terms of adapting the cartridge to alternate weapons platforms like belt-fed SAWs and especially its use in ultra-short barreled weapons tailored for CQB. The 6.5mm Grendel's capability with regards to terminal effect may also need to be further studied. A side by side comparison with the 6.8mm SPC (already tested in Afghanistan) would give the military a better idea of which cartridge is better suited as a possible replacement for the 5.56mm. An intermediate round, with half the recoil of the 7.62mm NATO, would certainly improve hit potential, especially at long engagement ranges. It's a shame that the U.S. military is facing this issue now, since the British had come to this conclusion more than sixty years ago. Insisting on a specific caliber battle cartridge for nostalgic reasons when supporting data showed other calibers to be more effective was a decision that made the U.S. look stubborn, overly prideful, and, frankly, just plain dumb. Apparently that is how Europe and much of the world viewed the decision at the time as well. The current U.S. consideration of intermediate cartridges that are almost identical in performance to the British .280 round of the late 1940s merely serves as a reminder of our mistake so many years ago.

Conclusion

Throughout the previous ten chapters, I have attempted to provide a brief history of how the United States has come to its present-day situation with respect to our current M16 series rifle and the increasing reliance on the old M14 for use in Afghanistan. It is rather ironic, but a mere generation ago many young U.S. soldiers were not even familiar with the M14. Most soldiers cannot say the same today. This is a rifle that only served for roughly seven years during its initial run as a standard service weapon. Those unfamiliar with the M14 would most likely consider the weapon a failure, given its short original service life; how wrong they are. Rather, the M14 was a victim of its time. The M14 served as a bridge between transforming views in service rifle design and use. It was a cross between earlier beliefs that a rifle should be made of wood and steel and have the power to reliably engage targets to 1,000 yards, along with the severe recoil generated, and the more modern view regarding the use of aluminum and plastic in rifle construction and the theory that dumping as many rounds downrange as possible in hopes of hitting the enemy would be just as effective as aimed semi-auto fire. This view turned out to be flawed, at least partially. The use of modern technology to save weight in the design of the M16 was a solid, forward thinking concept; it was just rushed into service without first perfecting the technology, which could have given us a reliable yet far lighter service weapon, and while the M16 proved light, it has proven to be far less reliable than the AK-47/AKM/AK-74 that it has traditionally faced.

As for the shortcomings of the 5.56mm cartridge itself, this was also a concept that was rushed into service. There were many variables involved in the M16 and its hurried adoption, in both the political and scientific arenas. The primary political issue stemmed from McNamara's dissatisfaction with the M14 rifle program and its cost overruns, in all probability. From a more technical standpoint, the 5.56mm did offer many advantages within its combat parameters; however, troops today are facing combat conditions that fall outside of those parameters and there's only so much the cartridge can be made to do.

As result of the U.S. caliber change so many years ago, our military finds itself once again evaluating intermediate rounds as it first did over sixty years ago. This was in hopes of finding the perfect rifle cartridge, something that does not and never will exist. While we may eventually see a new U.S. standard service rifle cartridge, many within the military have come to realize that effective fire for different shooting conditions requires different types of cartridges. There is no do-all solution, but our troops deserve the closest thing to perfection that we can get.

Regarding the absolute latest developments in intermediate cartridge technology, the 6.8mm SPC seems to be providing effective terminal capability and is likely the most fully developed combat cartridge under possible consideration. Its primary competitor, the 6.5mm Grendel, appears to offer superior long range performance and accuracy and is roughly equal in long range performance to the 7.62mm NATO. The 6.5mm Grendel offers this performance despite being lighter and shorter than the 7.62 NATO, as well as offering far less recoil. These cartridges are not without their flaws, however, and at the present time military interest in these cartridges seems to be waning. There are issues with both the 6.8mm SPC and 6.5mm Grendel concerning magazine performance in AR pattern weapons, in addition to concerns regarding their use in belt-fed automatic weapons. My guess is that the closest thing to an ideal intermediate cartridge for the military may have just appeared. There is a brand-new competitor that recently made its debut, the 6.5mm MPC (Multi Purpose Cartridge). This was introduced by SSK Industries.[1] This is the same company that developed the .300 Whisper, from which the .300 Blackout was developed. While these two cartridges (.300 BLK/6.5mm MPC) are roughly similar in size, it must be remembered that the 6.5mm bullet will likely offer far superior ballistic potential for long range shooting. This is a very new cartridge and far more evaluation will be needed regarding its performance, though it does offer some advantages over the other two contenders. Given its range of bullet weights and muzzle velocity, it compares favorably with the 6.5mm Grendel, though the Grendel was designed to handle somewhat heavier bullet weights. From existing data, it appears that the 6.5mm MPC has equal if not slightly better velocity than the Grendel, though the MPC has yet to see any factory loadings. With current handloading data, estimates for the 6.5mm MPC are a 95 grain bullet moving at 2,600 fps from a 14.5" barrel. This is compared to the 6.5mm

Grendel using a 110 grain bullet at about 2,500 fps.[2] This gives the MPC very similar ballistic potential. However, the MPC has the advantage of not having the same issues with magazine feeding and capacity as the 6.5mm Grendel or the 6.8mm SPC. It also can be used in belt-fed weapons without any changes to the belt links or feed mechanisms of current weapons like the M249 SAW. This is because the MPC has external case dimensions similar to those of the 5.56mm, where the 6.5mm Grendel and 6.8mm SPC have larger case diameters. In all, the new 6.5mm MPC does appear to offer the best low cost replacement for the 5.56mm, but as mentioned, this is a very new round and much work is left to be done regarding its development and evaluation.

One area of concern is that while the 6.5mm MPC offers superior long range performance to the .300 BLK, it does so with a lighter bullet, which is also smaller in diameter. It remains to be seen if the smaller bullet can provide the same level of terminal effect. One area of focus that will need to be addressed if the MPC is to compete with the .300 BLK is the development of a heavy bullet for a subsonic loading. If a satisfactory load cannot be developed, then the MPC will have no chance at reaching anywhere near the level of versatility offered by the .300 BLK. If such a load can be developed, then the U.S. may have come closer than ever before to finding an ideal infantry cartridge.

The question then becomes what platform from which to fire such a mythical round. While the U.S. has spent the last fifteen years developing a multitude of M16 accessories and adaptations, it has missed an opportunity to take new roads and push towards developing an ideal service weapon to go with such a cartridge. The fact that the Russians have developed small, light and hard hitting compact assault rifles proves that a new class of military weapon can always be developed. The ideal then would seem to have the following qualities. First, it should have an effective range superior to that of the 5.56mm NATO, as the limited ability of this round is a big reason for the writing of this book and the resurging popularity of battle rifles like the M14. It should have not only greater effective range than the 5.56mm but also superior terminal effect throughout this range. Second, use of the M16 pattern magazine would be a very cost effective measure, as this is supposed to be the NATO standard magazine, though it has not been accepted by Germany, as the G36 uses a different magazine. While the aluminum M16 magazine has not proven to be the most durable design over the years, the synthetic Magpul P-MAG has proven to be an excellent design and adds to the reliability of the M16 platform to no end. As there are many magazines of this pattern already in service, continued use would be a far less expensive option than developing a completely new design. Third, use of a proven conventional gas piston system would make for a more reliable overall weapon than what has been seen with the M16 series overall. Whether this takes the form of a short-stroke system like that used on the HK-416 or a long-stroke design similar to that of the AK-47 is not as important a matter. The recent PWS Mk 1 design has shown that a long-stroke AR pattern rifle is effective and, in fact, offers lower felt recoil. The use of a gas piston system also allows for less lubrication in keeping the weapon operating reliably. The M16 series requires a great deal of lubrication applied regularly in order to keep it up and running. This has proven to be a major problem in the sandy conditions found in the Middle East. Having a rifle that can continue to function with only a small amount of oil on the carrier would be a huge improvement and help greatly with keeping the sand out.

One area that would be a huge improvement over the M16/M4 series is the design of a good folding stock. A well-designed folding stock would allow for a major reduction in the overall length of the weapon. This had always been an issue with the M4. While the telescoping stock design of the recent M4 allows for adjustment and is a great benefit for troops wearing body armor, it only allows for a total reduction of three inches in overall length. This reduces the total length of the M4 from roughly thirty-three inches down to thirty. This is not the best length to work with when going indoors, especially if narrow hallways are involved. This is one area where the top folding stocks of the Russian compact rifles shine. The SR-3 and 9A-91 are both under sixteen inches long with the stocks folded. In designing something similar, a better folding stock layout would be a good idea, as top folding stocks are meant for reduction in total length and bulk and, while they do a good job here, they are not necessarily the best for strength and comfort. A well-designed side folding stock would be almost as compact and would provide for a much more user friendly rifle in terms of handling and natural pointing characteristics.

The ultimate goal in designing such a compact assault rifle would be to minimize weight and size while keeping the ability to engage targets out to most combat ranges. The ideal would be a weapon with a total empty weight of less than five pounds and an overall length of less than eighteen inches with the stock folded. This may seem like a lofty goal, but the Russian guns are capable of good terminal effect out to 450 yards with the heavy 9×39mm subsonic bullet. It is entirely within the realm of possibility given that the Russian weapons weigh less than four and one-half pounds in the empty condition.

Some standard features on such a weapon that would be a must are a

1913 rail to allow for a variety of sighting options along with an effective flash hider/muzzle brake that can take a quick detach suppressor. Another nice feature to consider would be a quick change barrel feature or one that could be changed out by an armorer at the unit level. This would allow a choice of barrel lengths to either extend the long range capability of the weapon or allow for a greater capability in the sustained fire role.

Cost could also be kept to a reasonable level, as it is not written in stone that the rifle use an expensive aluminum forging as the M16 does. Steel stampings are used on both Russian rifles and the weight is not an issue. With the current level of plastics use in firearms, weight in a well-designed rifle would be kept to a minimum. Careful thought would have to be applied here, however, as use of plastics in the wrong area can create drastic accuracy issues once the weapon heats up after heavy fire.

If such a rifle was to see adoption in the future, the resulting weapon would not have the extreme long range ability that is provided by the battle rifle, but it would be capable of accurate and lethal fire out to over 750 yards, with the ability to provide harassing fire out to over 900. It would be small and light enough to handle the role of a CQB weapon. In addition, it could accept a variety of sighting options. Most important, with the versatility of two separate combat loads it could maintain its close range lethality by using the heavy subsonic load in combination with a sound suppressor, while the lighter, high velocity load would provide the same lethality out to the ranges mentioned earlier. This would seem to be an enviable goal for the military, as it would help reduce the burden on the average trooper, as he already has to carry more than he wants into the field on a regular basis. My best guess is that such a project has been proposed more than once at the Pentagon, but given the current state of military involvement, it makes perfect sense to hold off until such a project can be managed without the risk of "working out the bugs" of a new weapon system in the field, where normal development problems could get U.S. troops killed. Enough have already died and it remains to be seen if Afghanistan will stabilize once they have gone home.

However, should the U.S. finally begin the process to find a replacement for the M16/M4, a small combat rifle laid out as described earlier would come closer than ever before to building the jack-of-all-trades service rifle. The reality has never changed when it comes to small arms design. No one single weapon can perfectly fill every role, but with the technological advancements in recent years we can come close.

This is all future possibility. As for the present, the current shakeup in the Middle East has all involved nations rather edgy, and due to this situation our military will no doubt take a conservative approach regarding the adoption of a new standard caliber at a rather touchy time. In other words, now is not the time for changing the current small arms/ammunition situation. Re-tooling would take too much time and money and the inevitable bugs will no doubt need to be ironed out first.

This brings us to the remaining portion of this work. This book was written with the intention of providing both a historical analysis for the current service rifle situation (7.62mm's return to popular use), and something of a technical descriptor for the 7.62mm NATO rifles that are out there today. While I most definitely excluded some lesser known models, or rather less prevalent models, by the same token I have tried to provide a solid rundown of the most common battle rifles. Given the last three rifles covered are quite new designs, the world unfortunately still sees a very strong future for the Cold War 7.62mm NATO and its long range, hard hitting bullet.

The advantage of using an intermediate cartridge for a service rifle lies in the reduction in recoil for the shooter. A battle rifle in 7.62 NATO has a great deal of recoil, and learning to shoot such a rifle accurately can take time and will take a good amount of practice. This is especially true if the trainee is not used to the recoil forces generated by something like a high powered rifle or a heavy gauge shotgun.

In comparison, the M16 produces relatively weak recoil, and having an intermediate cartridge that generates only slightly stronger recoil would be a big help in training. Many troops just plain do not like firing a rifle like the M14 due to the severe kick. The HK G3 is even worse in this regard due to its delayed-blowback method of operation.

Either the .300 BLK or 6.5mm MPC would be a good choice in this regard, as both rounds generate far lower recoil forces than the 7.62mm NATO. Not only would this allow for more accurate semi-auto fire; it also would allow for a faster recovery time for a second shot. These intermediate cartridges stick more closely to the "assault rifle" concept in allowing for far more effective and manageable automatic fire as well. Effective full-auto fire is all but impossible with most battle rifles, and this is most certainly true in the case of the M14 if it is issued with its original pattern stock. To be sure, the much heavier pistol grip stock of the M14 EBR is an improvement in providing for better control during automatic fire, but it is unlikely that it would be as effective as that coming from a rifle using an intermediate round.

This whole issue had been studied and resolved (or so many thought) years ago when the Germans first developed the StG 44.

As for the terminal effect of the .300 BLK, on paper only it is roughly equal to the 7.62×39mm Soviet cartridge used in the traditional AK-47/ AKM rifle. In fact, the 7.62×39mm actually has a slight edge in muzzle velocity while using a bullet that is nearly identical in weight. The Russian rifle pushes its bullet about 100 fps faster than that of the .300 BLK. The standard AK 123 grain bullet is leaving the muzzle at around 2,300 fps and the .300 BLK pushes its 125 grain bullet out of a similar barrel length at about 2,200 fps. Downrange, the performance appears to equal out, suggesting that the .300 BLK uses bullets with better drag characteristics.

However, subsonic cartridges are another story. The much heavier bullets used in the .300 BLK are moving at just below the speed of sound when they leave the muzzle. This is usually around 1,000 fps. The terminal effect is maintained by using bullet weights of up to 240 grains. By comparison, the Russian 9×39mm subsonic cartridges developed for the 9A-91 and AS families are moving a larger 250–260 grain bullet at roughly the same 1,000 fps. These rounds are reputed to have great terminal effect at close range. It is reasonable to assume that the .300 BLK subsonic rounds will have a similar terminal effect on targets. The .300 BLK offers a huge advantage here, as one weapon both can fire the lighter bullet at the higher velocity that would be required for most military operations and still be able to use the much heavier subsonic loads for CQB applications when needed. The 7.62×39mm AK cartridge has yet to see any standard subsonic variant, although I am sure some experimental cartridges have been made in the past for evaluation. If such rounds were made, they may not have proven satisfactory in performance, as the 9×39mm has been in use since the early to mid–'90s and appears to be gaining in popularity.

As the 7.62×39mm has gained a good reputation over the years as a fine stopping round, especially at close ranges, there is good reason to expect the same level of performance from the .300 BLK high velocity combat loads. Time will tell, as it appears that the .300 BLK is gaining in commercial popularity and it appears to be here to stick around for some time. As for the 6.5mm MPC, its longer range capability appears to be superior to the .300 BLK. This is due to the fact that the 6.5mm class of bullets has some of the best drag characteristics of any bullet diameter. Whether or not the terminal effect of this round can equal that of the .300 BLK will have to be determined in further testing. The lack of any factory produced loads for the 6.5mm MPC is a current problem, and of course, as mentioned, the lack of any superheavy bullets in this caliber is an issue that will have to be resolved if it is to have any hope of matching the .300 BLK down the road. Given that the bullet is smaller in diameter and lighter than standard .300 BLK bullets, it is not going to be any great surprise if the MPC does not quite match the

.300 BLK in terms of close range terminal effect. Bullet design can have a great deal to do with improving this to a considerable degree. Of course, downrange the situation will likely be different, as the 6.5mm MPC will retain greater energy at extended ranges, much like the 6.5mm Grendel, and it will do so without a reduction in magazine capacity, as with the Grendel. In either case, the performance of these cartridges will be a welcome boost to morale for U.S. troops, as many serving in Afghanistan have come to lose faith in the 5.56mm in any of its various bullet weights. The much heavier bullet weights offered by these new rounds will mean a definite increase in hitting power. One issue will be the need to retrain troops to handle the vastly different trajectory curves produced by these slower and heavier bullets. This will be most pronounced in the subsonic loadings, although these will likely be limited issue loads for use only when the operation requires stealth, as the subsonic loads are intended for use in combination with a sound suppressor.

The real advantage to the heavier bullets used in these new cartridge designs is that they are capable of providing lethal harassing fire to much greater ranges than can be achieved with the 5.56mm. No matter how good the bullet design is in terms of aerodynamic qualities, the much lighter .22 caliber bullets used in the M16 series bleed velocity at a much faster rate and lose energy in leaps and bounds. This has been the major complaint with the 5.56mm for years, and it is the reason that so many different bullet weights are in service, as the attempts to improve on this never seem to end. It began with the M193 55 grain bullet that all but died by 500 yards. The M855/SS109 62 grain load adopted in 1980 was a definite improvement in armor penetration and saw a smaller increase in effective range, with most agreeing that it tended to max out at 600 yards. More recent attempts led to the Mk262 series with its slower 77 grain bullet. This round has not been very impressive in terms of armor penetration, but its effective range has supposedly been increased to almost 750 yards. The most recent is of course the M855A1 with its fast burning powder and increased chamber pressure to help improve the short-barrel performance of the M4 series. To be fair, the new cartridge has improved the armor penetration ability of the 5.56mm by a considerable amount and its accuracy is reputed to be quite good. Its lethal range is still in question, though. All of this would be dealt with handily should a new heavy bullet intermediate cartridge see adoption.

One factor that has helped U.S. troops serving in Afghanistan has been the necessary training adaptations to better help them deal with the unique combat conditions they face. A very well-done military paper by U.S. Army major T.P. Ehrhart in 2009 may have had much to do with this. In his work, he outlines the major issues confronting troops currently serving there and his recommendations to help alleviate the problems. The major theme of the work was the shortcomings of the 5.56mm cartridge in general and the M855 round in particular. However, his work also concluded that the current U.S. Army rifle qualification program was in need of modification, as the one in use at the time did not provide theater specific training and the ranges at which troops were engaging were often beyond what troops had experienced in training. The U.S. Marine Corps still provides rifle qualification ranges to 500 yards, but the Army only takes the maximum range to 300 yards.[3] He also felt that troops should be better schooled in angular adjustments for shooting at distance. This was needed due to the highly rugged terrain and rapidly changing elevations present in Afghanistan.

Shortly after this paper's issue, the Army had modified its rifle qualification course to much better reflect the current combat conditions the troops were facing.[4] This is a remarkably fast move for a machine as large as the U.S. Army and it should help improve the ability of the units currently serving in those hills. In the Army's defense, there is only a limited amount of time and money allocated to the training of each individual soldier and the Army does have to prepare these men as best as possible to deal with a variety of combat conditions. The military had just altered its general scope of training to better reflect the likelihood of urban combat within the last two generations. Tradition has also often stalled training flexibility within the military over the years. It must be remembered that the U.S. had spent the larger part of the last seventy years fighting the Cold War and our military's focus on the probable combat conditions in the European theater had become institutionalized. The current downsizing and budgetary constraints in the U.S. military have all but forced them to learn to better adapt to changing conditions and times. This is a perfect example of the stereotype regarding a smaller yet more flexible military rather than a large yet ungainly service as maintained in the post-war years.

With regard to the long range marksmanship issues, or rather lack of, the Army has not traditionally placed the same emphasis on this aspect of rifle training for several reasons. Again, a large machine is not always the fastest off the line. The focus on closer ranges as the norm is due to the post-war trend towards urbanized combat (a more recent trend) in addition to the Cold War period's emphasis on expected European combat conditions. The budgetary constraints were a major factor as well.

The Marine Corps, however, never really relinquished their cherished value as riflemen. While this also was an example of military traditionalism, it happens to benefit the Marines in the current conditions in Afghanistan.

Without the proper training in long range marksmanship, a soldier can't possibly be expected to reliably hit a target at those longer distances. This is especially true for larger, slower cartridges, which tend to have a more pronounced trajectory in flight.

Even with a flat shooting round like the 5.56mm, this lack of long range shooting experience is going to be felt. At ranges within 300 meters (330 yards), the 5.56mm can be used to engage combatants by even those troops with minimal trigger time. This ability has been made even greater by the introduction of the 50/200m zero range for the M16/M4 series, as opposed to the earlier 300m zero of the M16A2 and the 250m zero of the M16A1. With this new zero range, the shooter doesn't have to alter point of aim or make sight adjustments within the first 300m of bullet travel.

However, once the shooting distances exceed this limit, even the flat shooting cartridges like the 5.56mm (also 5.45×39mm Soviet) begin to feel the inescapable effects of gravity. Without prior knowledge and experience regarding the behavior of a particular cartridge at those ranges, hits on targets are going to be inconsistent at best.

In Major Ehrhart's paper, he addresses the issue of the fairly new concept of the Designated Marksman specialty now present in the standard U.S. infantry squad. This position was created to offer reliable engagement options when ranges exceed those within the standard rifleman's ability. It is also suggested that this is not an entirely adequate solution to the problem regarding long distance engagement. While this approach to the issue has proven to have a valid purpose, this sort of long range engagement capability is needed on a larger scale to allow the squad to achieve its maximum potential level of accurate small arms fire.

The Squad Designated Marksman position has drawn a great deal of attention with regard to either new or modified weapons design. However, these are generally not widely issued weapons. The U.S. Marine Corps initially had about a thousand select M14s made up for this purpose.[5] This was deemed to be the least expensive solution when the M14s were compared to the rather pricey new weapons made for this job.

The British troops operating in the Middle East have taken a similar approach with the L129A1 sharpshooter rifle, and given the current production of so many new 7.62mm NATO service rifles it is becoming rather difficult to put forth a good argument that the battle rifle is dead. Quite the contrary, it is back, and to use a movie quote from an old Marine Corps Gunny, it is born again hard.

Chapter Notes

Chapter 1

1. Edward C. Ezell, *Small Arms of the World* (Harrisburg, PA: Stackpole Books, 1983), 19.
2. Ezell, 19.
3. Ezell, 19.
4. Ezell, 25.
5. Ezell, 29.
6. Norman Hitchman, "Operational Requirements for an Infantry Hand Weapon," Operations Research Office, Johns Hopkins University, 1952, 2 (Declassified).
7. Donald L. Hall, "An Effectiveness Study of the Infantry Rifle," Ballistic Research Laboratories, 11 (Declassified).
8. C.J. Chivers, *The Gun* (New York: Simon & Schuster, 2010), 276.
9. Larry Kahaner, *AK-47: The Weapon That Changed the Face of War* (Hoboken, NJ: John Wiley and Sons, 2007), 42.
10. Ezell, 47.
11. Chivers, 279–280.
12. "Report of the Rifle Review Panel," Department of the Army, 1968, C-6 (Declassified).
13. Chivers, 282.
14. Ezell, 47.
15. Ezell, 58.
16. Chivers, 281.
17. "Rifle Evaluation Study," U.S. Army Combat Developments Command, 1962, 5 (Declassified).
18. Chivers, 325.
19. Chivers, 327.
20. Kahaner, 228.
21. Major T. P. Ehrhart U.S. Army, "Taking Back the Infantry Half-Kilometer," 2009, 24–25.
22. Ibid.

Chapter 2

1. Ezell, 263.
2. Ezell, 514.
3. Ezell, 24.
4. Charles Cutshaw, *Tactical Small Arms of the 21st Century* (Iola, WI: Krause, 2006), 185.
5. "Belgium's FN FAL Battle Rifle," accessed April 20, 2013, http://www.armedforcesmuseum.com/beliums-fn-fal-battle-rifle.
6. Springfield Armory National Historic Site, Fact Sheet #4, "Post WW-II Rifle Development."
7. Ezell, 24.
8. Ezell, 20.
9. FN Herstal.
10. "DS Arms SA58 Para Tactical .308," accessed February 3, 2013, http://www.tactical-life.com/guns-and-weapons/ds-arms-sa58-short-spr-762mm/.
11. "Fusil Automatique Legere," accessed January 4, 2013, http://www.remtek.com/arms/fn/fal/index.htm.
12. DS Arms, Inc.
13. Ezell, 25.
14. "Britain's L129A1," accessed January

6, 2013, http://www.americanrifleman.org/m-articlepage.aspx?id=3221&cid=1.
15. Peter Kokalis, "FAL in Its Finest Form," *Soldier of Fortune* (October 2000), 44.
16. Kokalis, "FAL in Its Finest Form," 45.
17. DS Arms, Inc.
18. Ezell, 24.

Chapter 3

1. John Walter, *Modern Military Rifles* (London: Greenhill Books, 2001), 75.
2. HK-USA.
3. Ezell, 630.
4. "Rhodesian Bush War," accessed January 9, 2013, http://www.en.m.wikipdeia.org/wiki/Rhodesian_Bush_War.

Chapter 4

1. Lee Emerson, *M14 Rifle History and Development*, 2006, online edition accessed September 3, 2012, http://www.imageseek.com/m1a/M14_RHAD_Online_Edition_051010.pdf.
2. R. Blake Stevens, *U.S. Rifle M14: From John Garand to the M21*, 2nd Ed. (Cobourg, Ontario, Canada: Collector Grade Publications, 1991), 175.
3. Heckler & Koch G3, accessed September 7, 2012, http://www.bazpedia.com/en/h/e/c/Heckler_%26_Koch_G3_4125.html.
4. Ezell, 25–26.
5. Ezell, 26.
6. Emerson, 35–36.
7. Emerson, 26.
8. Ezell, 26.
9. Ezell, 764.
10. Ezell, 155.
11. Emerson, 74.
12. Smith Enterprises, Inc.
13. Emerson, 61.
14. Ezell, 16.

Chapter 5

1. Ezell, 33.
2. Ezell, 33.
3. Walter, 108.
4. Walter, 108.
5. Ian Hogg, *Military Small Arms of the 20th Century*, 7th Ed. (Iola, WI: Krause, 2000), 283.
6. Cutshaw, 191.
7. FN FAL, accessed November 15, 2013, http://en.wikipedia.org/wiki/Bren_light_machine_gun.
8. Fábricas y Maestranzas del Ejército de Chile, accessed February 8, 2013, http://www.famae.cl.
9. Ian Hogg, ed. *Jane's Infantry Weapons 1993–94*, 19th Ed. (Alexandria, VA: Jane's Publishing Group, 1993), 195.
10. French RFP, accessed February 17, 2013, http://www.thefirearmblog.com/blog/2011/11/23/french-army-to-replace-the-famas=rifle/.
11. Walter, 110.
12. SIG 550/551 Test Data and Documentation, accessed February 17, 2013, http://www.bighammer.net/sigamt/550/550techinspection/.
13. Ehrhart, 44.

Chapter 6

1. Chivers, 12.
2. Ezell, 552.
3. Yom Kippur War, accessed February 17, 2013, http://www.israeli-weapons.com/history/yom_kippur_war/yomkippurwar.html.
4. Cutshaw, 209.
5. Cutshaw, 211.
6. IMI Galil, accessed February 3, 2013, http://www.en.m.wikipedia.org/wiki/IMI_Galil.
7. Walter, 87.
8. Galil, accessed February 3, 2013, http://www.remtek.com/arms/imi/galil/galil.htm.
9. Cutshaw, 272.
10. Indumil, http://www.Indumil.gov.co.
11. IWI, http://www.israel-weapon.com.

Chapter 7

1. Walter, 35.
2. Cutshaw, 253.

3. Cutshaw, 253.
4. FN Herstal, http://www.fnherstal.com.
5. FN Herstal, http://www.fnherstal.com.
6. FN Herstal, http://www.fnherstal.com.
7. FN Herstal, http://www.fnherstal.com.
8. "SAS to Use Bigger Bullets to Kill Enemy Outright after Claiming 'Shoot-to-Wound' Policy Put Their Lives at Risk," accessed February 18, 2013, http://www.dailymail.co.uk/news/article-2294631/SAS-use-bigger-bullets-kill-enemy-outright-shoot-wound-policy-lives-risk.html?_mchannel=rss&ns_campaign=1490.

Chapter 8

1. Walter, 82.
2. "M4 Carbine Controversy," accessed February 12, 2013, http://www.defenseindustrydaily.com/the-usas-m4-carbine-controversy-03289/#more-3289.
3. "Face Time with the HK 416—The Gun That Killed Bin Laden," accessed March 3, 2013, http://www.popularmechanics.com/_mobile/technology/militaryweapons/face-time-with-the-hk416-the-gun-that-killed-bin-laden#slide-1.
4. "It's Better Than the M4, But You Can't Have One," accessed February 12, 2013, http://www.airforcetimes.com/article/20070219/News/702190356/It-s-better-than-M4-you-can-t-one.
5. Heckler & Koch, 2012.
6. Heckler & Koch, 2012.
7. Heckler & Koch, 2012.
8. G28 Marksman Rifles, accessed March 12, 2013, http://www.shootingillustrated.com/mobile/blog.php?id=19445&cid=858.

Chapter 9

1. IWI Galil ACE, accessed February 18, 2013, http://www.webcitation.org/5s3iMywSv.
2. Galil ACE, accessed February 12, 2013, http://www.world.guns.ru/assault/isr/galil-ace-e.html.
3. Galil ACE, accessed February 12, 2013, http://www.en.m.wikipedia.org/wiki/Galil_ACE.
4. "Galil ACE Rifle Adopted by Guatemala National Civil Police," accessed February 12, 2013, http://www.thefirearmblog.com/blog/2011/02/16/iwi-galil-ace-rifle-adopted-by-guatemala-national-civil-police/.
5. IWI, 2012.
6. Galil, accessed February 12, 2013, http://www.strategypage.com/htmw/htproc/articles/20100128.aspx.
7. IWI, 2012.

Chapter 10

1. History of Gunpowder, accessed February 12, 2013, http://www.en.m.wikipedia.org/wiki/History_of_gunpowder.
2. Percussion cap, accessed February 12, 2013, http://www.en.m.wikipedia.org/wiki/Percussion_cap#References.
3. The Reverend A. J. Forsyth, accessed February 12, 2013, http://www.en.m.wikipedia.org/wiki/Percussion_cap#References.
4. Pinfire System, accessed February 12, 2013, http://www.en.m.wikipedia.org/wiki/Pinfire.
5. Frank C. Barnes, *Cartridges of the World*, 13th Ed. (Iola, WI: Krause, 2012), 494.
6. Barnes, 11.
7. Barnes, 96.
8. Barnes, 365.
9. Barnes, 157.
10. Barnes, 157.
11. Smokeless powder, accessed February 16, 2013, http://www.en.wikipedia.org/wiki/Smokeless_powder.
12. Corrosive ammunition, accessed February 16, 2013, http://www.shootingillustrated.com/index.php/6372/a-case-for-corrosive-ammo/.
13. Hogg, 286.
14. History of M118 ammunition, accessed February 18, 2013, http://www.snipercentral.com/M118.phtml.
15. 7.62mm NATO, accessed February 18, 2013, http://www.en.wikipedia.org/wiki/7.62%C3%975/mm_NATO#Military_cartridge_types.
16. 7.62x51mm NATO, accessed February 18, 2013, http://www.en.wikipedia.org/wiki/7.62X51mm_NATO.

17. 7.62mm (7.62x51mm), accessed February 18, 2013, http://www.inetres.com/gp/military/infantry/rifle/762mm_ammo.html.

18. "The Longest 7.62 Kill Shot," accessed February 18, 2013, http://www.tactical-life.com/tactical-weapons/the-longest-762-kill-shot/2/.

19. U.S. Army FM-3-22-9, Table 2-1.

20. 7.62X51 NATO (.308 Win.), accessed February 18, 2013, http://www.snipercentral.com/308.htm.

21. Ezell, 62.

22. Ehrhart, 56.

23. U.S. Army TM-9-1005-249-10, 38.

24. Barnes, 374.

25. Barnes, 374.

26. "The Last 'Big Lie' of Vietnam Kills U.S. Soldiers in Iraq," accessed February 18, 2013, http://www.americanthinker.com/2004/08/the_last_big_lie_of_vietnam_ki.html.

27. United States Marine Corps Scout Sniper, accessed February 18, 2013, http://www.en.m.wikipedia.org/wiki/Scout_Sniper.

28. "Effective Range for Colt M4 vs. Mk 12 Mod. 0," accessed February 18, 2013, http://www.M4carbine.net/showthread.php?p=313273.

29. "5.56x45mm NATO," accessed February 18, 2013, http://www.gunsandammo.org/index.php?title=5.56x45mm_NATO.

30. "The Last 'Big Lie.'"

31. Hogg, 223.

32. "Accurate AR Fodder," accessed February 18, 2013, http://www.shootingillustrated.com/mobile/article.php?id=8072.

33. "Evolution of the M85A1 Enhanced Performance Round," accessed February 18, 2013, http://www.army.mil/article/48657/.

34. "Army Won't Field Deadlier Corps Round," accessed February 20, 2013, http://www.armytimes.com/article/20100402/NEWS/4020308/Army-won-t-field-deadlier-Corps-round.

35. "The Last 'Big Lie.'"

36. "In Praise of the M16 Rifle," accessed February 20, 2013, http://www.tactical-life.com/special-weapons/in-praise-of-the-M16-rifle/.

37. Ehrhart, 23.

38. "Reviewing Black Hills Mk 262 Mod 1 Ammo," accessed February 20, 2013, http://www.shootingtimes.com/2012/03/21/special-forces-to-civilians-black-hills-mk-262-mod-1-review/.

39. "M855A1: Should It Be the New Round for Soldiers and Marines," accessed February 20, 2013, http://www.gunsandammo.com/2012/03/07/m855a1-should-it-be-the-new-round-for-soldiers-and-marines/.

40. Ezell, 715.

41. Steve Crawford, *Deadly Fighting Skills of the World* (London: Macmillan, 1997), 109.

42. Barnes, 385.

43. Crawford, 101.

44. AK-47, accessed February 20, 2013, http://www.en.m.wikipedia.org/wiki/AK-47.

45. 2B14 Podnos, accessed February 21, 2013, http://www.en.m.wikipedia.org/wiki/2B14_Podnos.

46. "Some Problems in the Field with the Mk-17 SCAR," accessed February 21, 2013, http://www.kitup.military.com/2011/01/some-problems-in-the-field-with-mk-17-scar.html.

47. KPV, accessed February 21, 2013, http://www.en.m.wikipedia.org/wiki/KPV.

48. "Soldiers Benefit from Lighter, Easier to Maintain Mortar Systems," accessed February 22, 2013, http://www.army.mil/article/61843/Soldiers_benefit_from_lighter_easier_to_maintain_mortar_systems/.

49. 6.8mm SPC Cartridge History & Development. Hornady's Ammunition. The Stag Carbine, accessed February 22, 2013, http://www.demigodllc.com/articles/6.8-mm-spc-cartridge-history-development-hornady-stag-arms-carbine/?p=6.

50. Ehrhart, 31.

51. See note 49.

52. 6.5mm Grendel, accessed February 22, 2013, http://www.en.m.wikipedia.org/wiki/6.5mm_Grendel.

53. "6.5mm Grendel Gets Official: SAAMI Specs, Saigas, and Veprs," accessed February 22, 2013, http://www.guns.com/2011/10/19/65mm-grendel-gets-official-saami-specs-saigas-and-veprs/.

54. "Tactical Titans," accessed February

22, 2013, http://www.rifleshootermag.com/2010/09/23/featured_rifles_rs_tactical titans_200906/.
55. "6.5 Grendel," accessed February 22, 2013, http://www.shootingillustrated.com/mobile/article.php?id=1759.
56. Introduction: The .222 Remington, accessed February 22, 2013, http://www.6mmbr.com/compcartridges.html.
57. Russia, Assault Rifles, accessed February 22, 2013, http://www.world.guns.ru/assault/rus/cp-3-cp-3M-vortex-e.html.
58. Cutshaw, 225.
59. Izhmash AK-9 Assault Rifle, accessed February 22, 2013, http://www.militaryfactory.com/smallarms/detail.asp?smallarms_id=579.
60. Advanced Armament Corporation, 2012.
61. SSK Industries, 2012.
62. Advanced Armament Corporation, 2012.
63. 6.8×43mm SPC (Special Purpose Cartridge), accessed February 18, 2013, http://www.globalsecurity.org/military/systems/munitions/68spc.htm.
64. AR-15 Lower Receivers, accessed February 16, 2013, http://www.surplusammo.com/categories/AR%252d15-Receivers/Lower-Reveivers/.
65. Rifle Recoil Table, accessed February 18, 2013, http://www.chuckhawks.com/recoil_table.htm.
66. Shepherd Enterprises, Inc.
67. Alexander Industries, Inc.
68. Ibid.
69. "M855A1: Should It Be the New Round."
70. 5.56mm (5.56×45mm) Ammunition, accessed February 18, 2013, http://www.inetres.com/gp/military/infantry/rifle/556mm_ammo.html.
71. Ehrhart, 33.
72. 7.62mm Penetration Question, accessed February 18, 2013, http://www.leatherneck.com/forums/showthread.php?15919-7-62mm-penetration-question.
73. M855A1, accessed February 18, 2013, http://www.usarmorment.com/pdf/M855A1.pdf.
74. "M855A1: Should It Be the New Round."

Conclusion

1. SSK Industries.
2. "SSK 6.5 MPC: Best Assault Rifle Cartridge for 21st Century Warfare?" accessed March 3, 2013, http://www.defensereview.com/ssk-65-mpc-best-assault-rifle-cartridge-for-21st-century-warfare/.
3. Ehrhart, 53.
4. "Weapons Training and Qualification Overhauled," accessed March 15, 2013, http://www.armytimes.com/article/20080504/NEWS/805040329/Weapons-training-qualification-overhauled.
5. Jack Lewis, *The Gun Digest Book of Assault Weapons,* 6th Ed. (Iola, WI: Krause, 2004), 94.

Bibliography

Books

Barnes, Frank C. *Cartridges of the World,* 13th Ed. Iola, WI: Krause, 2012.
Chivers, C.J. *The Gun.* New York: Simon & Schuster, 2010.
Crawford, Steve. *Deadly Fighting Skills of the World.* London: Macmillan, 1997.
Cutshaw, Charles. *Tactical Small Arms of the 21st Century.* Iola, WI: Krause, 2006.
Emerson, Lee. *M14 Rifle/History and Development*, online edition. 2006.
Ezell, Edward C. *Small Arms of the World.* Harrisburg, PA: Stackpole Books, 1983.
Hogg, Ian, ed. *Jane's Infantry Weapons, 1993–94.* 19th Ed. Alexandria, VA: Jane's Publishing Group, 1993.
Hogg, Ian. *Military Small Arms of the 20th Century,* 7th Ed. Iola, WI: Krause, 2000.
Kahaner, Larry. *AK-47: The Weapon That Changed the Face of War.* Hoboken, NJ: John Wiley and Sons, 2007.
Lewis, Jack. *The Gun Digest Book of Assault Weapons,* 6th Ed. Iola, WI: Krause, 2004.

U.S. Government Documents

Department of the Army. "Report of the Rifle Review Panel." 1968, C-6 (Declassified).
Ehrhart, T.P. "Taking Back the Infantry Half-Kilometer," 2009.
Hall, Donald L. "An Effectiveness Study of the Infantry Rifle." Ballistic Research Laboratories, 11 (Declassified).
Hitchman, Norman. "Operational Requirements for an Infantry Hand Weapon." Operations Research Office, Johns Hopkins University, 1952, 2 (Declassified).
Springfield Armory National Historic Site. "Post WW-II Rifle Development." Fact Sheet #4.
U.S. Army, FM-3-22-9.
U.S. Army, TM-9-1005-249-10.
U.S. Army Combat Developments Command. "Rifle Evaluation Study." U.S. Army Developments Command, 1962, C-6 (Declassified).

Periodicals

Soldier of Fortune, October 2000.

Electronic Sources

Accurate Shooter.com, 6mmbr.com
Air Force Times, airforcetimes.com
American Rifleman, americanrifleman.org
American Thinker, americanthinker.com
Armed Forces History Museum, armedforcesmuseum.com
Army Times, armytimes.com
Big Hammer.net
Gary W. Cooke, inetres.com
D.E. Israeli Weapons Ltd., israeli-weapons.com
The Daily Mail, dailymail.co.uk
Defense Industry Daily LLC, defenseindustrydaily.com
Defense Review, defensereview.com
DEMIGOD, LLC, demigodllc.com
Fabricas y Maestranzas del Ejercito, famae.cl
FN Herstal, S.A., fnherstal.com
globalsecurity.org
Guns & Shooting Online, chuckhawks.com
guns.com
Guns and Ammo, gunsandammo.org
Industria Militar de Colombia, Indumil.gov.co
Israeli Weapon Industries (IWI) LTD, israel-weapon.com
Leatherneck.com
Lee Emerson webpage, imageseek.com
M4carbine.net
militaryfactory.com
Modern Firearms & Ammunition, world.gun.ru
Monster Worldwide, Inc., kitup.military.com
Popular Mechanics, popularmechanics.com
REMTEK, the ARMS site, remtek.com
Rifle Shooter Magazine, Rifleshootersmag.com
Shooting Illustrated, Shootingillustrated.com
Shooting Times, shootingtimes.com
Sniper Central LLC, snipercentral.com
Strategy World.com, strategypage.com
Surplus Ammo & Arms LLC, surplusammo.com
Tactical-Life, tactical-life.com
thefirearmblog.com
United States Army, army.mil
U.S. Armorment LLC, usarmorment.com

Manufacturing Sources

Advanced Armament Corporation
Alexander Industries, Inc.
DS Arms, Inc.
FN Herstal, S.A.
Heckler & Koch GmbH
Israel Weapon Industries (IWI), Ltd.
Shepherd Enterprises, Inc.
SSK Industries

Index

.22 7–8, 101, 155, 166
.22 rimfire 101
.220 Russian 142
.222 Remington 8
.222 Remington Special 8
.223 Remington 8
.270 5
.276 Pedersen 5, 126
.280 6–7, 18, 158
.30 calibers 14, 133, 156
.30 Remington 146
.30–06 6, 19, 103, 108, 126–27, 130, 132–34, 136
.30–30 (.30 WCF) 104, 108
.300 BLK (7.62×35mm) 150–55, 160–61, 164–66
.300 Whisper 150, 160
.303 5, 143–44
.308 Winchester 6, 133
.357 Magnum 103
.44 rimfire 105–06
.44–40 102, 106–07
.44–77 102
.45 ACP 133, 149
.45 LC 102, 106
.45–60 107
.45–70 Government 102, 107
.45–75 107
.50 Beowulf 155
.50 BMG 103
.50 Whispers 150
.50–90 102, 107
.50–95 Express 107
.50–110 Express 107
12.7mm 144
12S (Beretta) submachine gun 26

14.5mm 144
1860 Henry rifle 104–06
1873 SAA "Peacemaker" 102
1873 Winchester 102, 104, 106–07
1874 Sharps 102, 107
1876 Winchester 104, 107
1886 Winchester 104, 107
1888 Lee-Metford 111
1891/ 91/30 Mosin-Nagant 111, 113, 117–18, 122
1892 Krag 109
1892 Winchester 107
1894 Winchester 104, 108
1895 Winchester 108–09
5.45×39mm Soviet 12–13, 142, 157, 168
5.56mm NATO (5.56×45mm) 8–9, 12–15, 29, 33, 39, 43, 45, 48, 62, 65–68, 72–77, 79–80, 84, 86–88, 95, 97, 136–47, 151–61, 166–68
50.61 21
50.63 carbine 21, 28, 66
50.64 21
50/200 zero 168
510/510–1/510–2/510–3/510–4(SIG) 43, 61–62
540–1/543–1/542–1 62–64, 67–68, 97
5N7 13
6.5mm Grendel (6.5×38mm) 147–48, 151–56, 158, 160, 166
6.5mm MPC 160–61, 164–66
6.8mm SPC (6.8×43mm) 146–48, 151–56, 158, 160–61
60mm mortar 145
63 series *see* Stoner model 63
6th Army (Germany) 123
7.5×55mm 61

Index

7.62mm NATO (7.62×51mm) 6–8, 10, 12, 14–15, 18–20, 29–31, 33, 42–43, 52, 54, 61–68, 71–77, 81–82, 84, 86–90, 95, 97–98, 100, 111, 126, 132–38, 141–45, 147–48, 153–58, 160, 164, 168
7.62×25mm Tokarev 117
7.62×39mm Soviet 12, 26, 33, 43, 62, 75, 95, 142–43, 146, 148, 151, 165
7.62×54mmR 109, 142–144
7.92×33mm 6, 16, 126
75th Rangers 80
81mm mortar 145
82mm mortar 143, 145
9A91 149
9×19mm Parabellum (9mm Luger) 66, 71, 114, 120, 122, 125, 133, 149
9×39mm 149–50, 162, 165

Aberdeen Proving Grounds 7
Acetone 35
ACOG 154
Advanced Armament Corporation (AAC) 150
Afghanistan 2, 13, 14–15, 17, 29–30, 45–46, 48, 58, 60, 83, 94, 137–138, 140–45, 147, 151, 153–54, 156, 158–59, 163, 166–67
Africa 17, 23, 31, 33, 47, 62, 68, 110, 130
AGILE 9–10
AK-100 150
AK-47/AKS-47 1–3, 7, 10, 12–13, 16, 24, 28, 33, 39, 44, 50, 58, 62, 64, 67–72, 74–76, 95, 98, 107, 110, 125, 142–43, 145–46, 148, 159, 162, 165
AK-74/AKS-74 12–14, 44, 142, 159
AK-74M 13
AK-9 150
AKM/AKMS 12–13, 33, 39, 44, 142–43, 145, 159, 165
AKSU-74 (AKS-74U) 13
Alexander, Bill 148
Alexander Arms 148
AM55 61
American Indian 19
American West 107
Ammonia 36
AR 70 (Beretta) 13
AR-15/ M16 8, 10–14, 24, 32, 34, 41, 45, 58, 68, 70–72, 77, 80–84, 87–88, 90–95, 97, 137– 41, 143–44, 146–48, 151–52, 156–57, 159, 161–64, 166, 168
AR-18 87
Arapaho 107
Argentina 16, 33

Arisaka 109
Armalite 8, 87
Armor piercing ammunition 14, 134–35, 137, 155
Armor piercing incendiary (API) 134
Army Ordnance 7, 54, 127
ARPA 9
AS 149
Assault rifle 1–3, 5, 7, 16, 18–19, 61, 66, 79, 114, 117, 124, 126, 136, 149, 161–62, 164
ATTD 31
AUG (Steyr) 13, 82, 86
Australia 16, 20, 22
Austria 13, 16, 20, 30
Auto-5 (Browning) shotgun 104
AVS-36 Simonov 113

Ball powder 11
Ballistic coefficient (BC) 146, 148, 150
Bang gas trap system 124, 127
"Bang-bang, jam" 20
BATFE 30
Battle of Kursk 124
Battle of Stalingrad 123
Battle of the Little Bighorn 107
BB cap 101
Bearing grease 37, 57, 92
Beeching, Richard, Dr. 5
Berlin 117
Black powder (gunpowder) 100, 102, 105, 107–08, 111
Blank round 134
"Blitzkrieg" 114, 116
BM 59 (Beretta) 54
Boutelle, Richard 8
Brass catcher 47
Brazil 16, 30–31, 34, 79, 97
Bren light machine gun 64, 132, 148
Brennan, Arne 148
Brest 121
Browning, John Moses 16, 104, 107–08, 119, 130–31
Browning automatic rifle (BAR) 1, 19, 112, 130
Buffalo (American bison) 102, 107
Bullpup 6, 18, 29, 56, 68, 72, 97, 125

C8 (Diemaco) 86
C96 Mauser "broomhandle" 112
CAL (FN) 79
Canada 16, 20, 86
"Carrier tilts" 87
Case head separation 11
Century International Arms, Inc. 45

CETME 6–7, 126
Chauchat light machine gun 132
Chervenak, Michael 11
Cheyenne 107
"Chicago typewriter" 112
Chile 62, 64, 97
Chosin Reservoir 28
Chrome plating 11, 36, 50, 64, 73, 78, 82
Civil War 105
Claw mount (HK) 40, 47
Cold War 1–2, 7, 17, 61, 133, 141, 164, 167
Colombia 72, 75, 95–98
Colt, Samuel 113
Colt 8–9, 72, 86, 102, 105, 106–07
Communist China 53
Cooper-MacDonald 9–10
Corrosive primer/ammunition 35, 103, 129

Delta Force 87
Diemaco 86
Dien Bien Phu 8
Direct impingement 24, 81, 88, 93–94, 144
DP/DPM Degtyaryov light machine gun 118, 122
DS Arms 30–31, 46
Dummy round 134

Eastern Front 115, 121
Effective range 2, 5, 39–40, 45–46, 52, 55, 58, 71, 73–74, 80, 117, 135–37, 139, 143–45, 151, 154–55, 161, 166
Ehrhart, T.P., Maj. (U.S. Army) 146, 166, 168
EM2 6, 18
Emerson 3
"en-bloc" clip 127
Enfield 6, 29, 109–10, 143

Fabrique Nationale (FN) 6–7, 16–19, 21, 23, 25, 27, 29, 31, 33, 35, 37, 66, 71, 79–81, 83–86, 88, 96, 125, 132, 143
Failure to extract (FTE) 11
Fairchild Corporation 8
FAL (FN) 3, 6–7, 14, 16–39, 41, 43–44, 46–49, 51–52, 57, 63–65, 71, 79, 86, 91, 97, 102, 104, 107, 111, 119, 122, 125–26, 137, 145, 160
FAL heavy barrel 20, 52
Falkland Islands 33
FAMAE 62, 63, 68, 69
FAMAS 13, 68–69
Federal 168 grain OTM (open tip match) 134
FG-42 19, 58, 114

Finland 72, 76, 113, 123, 125, 148
Flintlock 101
FNC (FN) 79
"Frontier Six-Shooter" 107
Forsyth, Alexander J., Rev. 101
Fort Benning, GA 6
Fulton Armory 53

G.258(r) 121
G.259(r) 121
G1 7
G28 (HK) 94
G3 7, 14, 17, 22–24, 29, 32–33, 38–52, 57, 64, 87, 89, 97, 109, 126, 161, 164
G3/G3A3/G3A4/G3A3Z/G3A3ZF/G3-SG1 (HK) 39–41, 45, 46, 109
G36 (HK) 45, 48, 87, 161
G36 (HK) 45, 48, 87, 161
G41 (HK) 87
G41 (m) 116–18, 121, 124
G41 (w) 116–17
G43 114, 116–18, 123–24, 126, 130
Galil ACE 75, 95–99
Galil ACE 31/32 95
Galil ACE 52/52L/53 95, 97
Galil ACE N 95
Galil ARM/AR/SAR 23–24, 44, 63, 72–78, 95–99
Galil Sniper rifle 75, 97
Gasoline 35
Gatling, Richard, Dr. 19
Graphite 27
Great Depression 103
Greece 17
Grenada 141
G-series FAL 32
Guatemala 95–96
Gulf War 48

Hague accords 10, 157
Hair tonic oil 28
Hallock, Richard, Col. (U.S. Army) 10
Harrington & Richardson (H&R) 18, 53
Heckler & Koch (HK) 6–7, 13, 17, 22–24, 29, 33, 38–50, 61, 64, 72, 87–94, 162, 164
High Standard 18
Hi-Power 16
Hitler, Adolf 115, 118, 123
HK 21 machine gun 87
HK 23 machine gun 87
HK 33 13, 87
HK 416 87–89, 162
HK 417 88–91, 93–94

Holland sight 23
Howard, Edward C. 101

IMBEL 30–31
"Inch pattern" vs. metric FAL 17, 34
India 17
Indumil 95, 98
Iran 15, 47
Iron sights 23, 40, 52, 89, 96, 135–37, 154
Israel 15, 20, 23–24, 26–28, 44, 56, 71–76, 95, 97
Italy 54

Kennedy, J. F., U.S. President 8, 10–11, 139
Kerosene 35
King's loading gate 106
Knight's Armament 82
Konev, Marshal 118
Korean War 7, 24–25, 28, 111–12, 131, 133
KPV heavy machine gun 144
Krag-Jorgensen 109

L129A1 3, 14, 30, 168
L1A1 Self Loading Rifle (SLR) 2, 19–21, 29–30, 33–34, 64, 111
L42A1 111
L4A1 light machine gun 64
L85 (SA80) 6, 29
Lapua 148
Latin America 63, 75, 96, 98
Leatherwood ART scope 55
Leatherwood mounts 23
Lemay, Curtiss, Gen. (U.S.A.F.) 8
Lewis light machine gun 120
Lewis Machine & Tool (LMT) 14
Libbey-Owens-Ford 129
Long-stroke gas piston 24–25, 28, 54, 65, 69, 72, 79, 95, 129, 162
Lubriplate 28
Lyman Alaskan scope 129

m/62 (Valmet) 72, 76
m/961 45
M1 Garand 1–2, 6, 9, 28, 52–54, 58, 113–14, 118, 121, 124, 126, 129–30, 133–134
"M1 thumb" 128
M1/M2 carbine 8, 24–25, 28, 54, 82
M118 134
M118LR 134, 154
M14 2–3, 6–8, 10–11, 14, 18–19, 22–23, 30, 36–37, 43–45, 47, 51–60, 80, 82, 92, 109, 129, 132, 134–36, 144–45, 153–54, 157, 159, 161, 164, 168

M14A1 19, 52
M15 52
M16A1 2, 12, 137, 139, 141, 168
M16A2 9, 12, 137, 139, 141, 168
M16A4 144–145
M1903/M1903A3/M1903A4 109, 135
M1913 "Picatinny" rail 47–48, 68, 88, 95
M1917 machine gun 19, 130
M1918/1918A1/1918A2 130–31
M1921 Thompson submachine gun 112
M193 9, 12–13, 137–39, 141, 157, 166
M1941 Johnson rifle 127
M1C/M1D 128–29
M2 (ball) 8, 134
M21 109
M224/224A1 144–45
M240 machine gun 144
M249 30, 144, 161
M4/M4A1 14, 31, 46–47, 52, 58, 66, 72, 80, 83, 89, 96–97, 137–38, 140, 143–46, 156–57, 162–63, 166, 168
M40 138
M43 142–43
M59 134
M60E1 machine gun 143
M80 134–35, 137, 145, 148, 156–57
M80A1 145, 156–57
M81/M82/M84 scope 129
M855 9, 14, 137–41, 143, 145, 155–57, 166–67
M855A1 EPR (Enhanced Performance Round) 138, 140, 142, 145, 152, 155–58, 166
M9 (Beretta) 112
M98/98k 12, 109–10, 113, 116, 118–21, 125
M993 135, 155–56
M995 156
MacArthur, Douglas, Gen. (U.S. Army) 126
Madsen light machine gun 1, 120, 132
MAG (FN) machine gun 97
Malinovsky, Marshal 118
MANURHIN 62
Marlin 108
Matchlock 101
Mauser 6–7, 12, 38, 109–12, 116, 119, 121, 125
Mauser, Paul 109, 119
Mauser, Peter 109
Mauserwerke, Oberndorf 6
Maxim machine gun 1, 112, 130
McNamara, Robert 9–11, 139, 159
"Meat axe" 9
MG08 (Maxim) machine gun 19
MG15 Bergmann light machine gun 120

Index

MG-34 116
MG-42 116
Middle East 15, 30, 67, 138, 147, 162–63
Mk 13 EGLM (FN) 80
Mk 14 Mod 0 (M14 EBR) 54, 56, 164
Mk 18 CQBR 80
Mk 20 SSR 82–84
Mk 262 Mod 1 138, 146, 152, 157
Mk 318 SOST (Special Operations Science and Technology) 142, 155
Mk III SMLE 111, 113
Mle 1930 (FN) 132
Mle D (FN) 132
Model 49 (SAFN) 16
Model 58 7
Mondragon rifle 112
Mongols 100–01
MP 18 Bergmann submachine gun 120
MP-40 114, 122, 125
MP-43 118, 125
MP-44 125
MP5 (HK) submachine gun 40–41, 87
MR762A1 (HK) 89–91, 94

Nagant brother 111
Native Americans 141
NATO 6-7, 10, 12–14, 18, 29, 37, 79–80, 95, 133, 139, 147, 161
Nazi Germany 19, 125
Negev NG7 97
Netherlands 20
NFA 30
Night sights 75, 78
No. 4 Mk 1 113
Norinco 53
Normandy 130
North Africa 130
North America 108
Norway 87, 109

Operation Barbarossa 113, 118, 121, 125
Operation Desert Storm 141
"Operational Requirements for an Infantry Hand Weapon" 7
Operational Specialist Weapon (OSW) 31
Operations Research Office (ORO) 7
Osama bin Laden 87
Ottoman Turks 109, 111

P-210 (SIG) 66
P220 (SIG) 66
P9/P9S (HK) 87
Pacific Theater 130, 135
Pakistan 43, 47, 97, 145

Panama 141
Patton, George S., Gen. (U.S. Army) 130
PE, PU scope 111
Percussion cap 101
PK/PKM machine gun 142–44
Plains Indians 102
P-Mag (Magpul) 70, 161
Pohjoispaa, Janne 148
"Poison bullet" 13
Polytech 53
Portugal 45
PPD-40 125
PPS-43 122
PPSh-41 113–14, 117–18, 122, 125
Primary extraction 12, 29
Primary Weapon Systems (PWS) Mk1 162
PTR-91 42, 45–47

R4 23–24, 68
Ramadi, Iraq 138
Rate of twist 12, 139
Remington 8, 107, 138, 146, 148
Rhodesian Bush War 48
Rifle grenade 64, 134
Rock Island Arsenal 10
Rosario 33
Rotary magazine 109, 127
RPG-7 143
Ruptured case extractor (stuck case puller) 11
Russian Federation 13

SA58, SA58 Congo 30
SAAMI 148
Saive, Dieudonne 16, 35
SALVO 7
Saudi Arabia 47
Savage 95, 99 108–09
SCAR-H/Mk17/Mk17S 66, 79–85, 88–89, 96
SCAR-L/Mk16/Mk16S 80, 83–84
Seal Team Six 87
Self Loading Rifle (SLR) 20
SG 530 62, 69
SG 540/542/543 (SIG) 13, 62, 64–66, 68–69, 76
SG 550/551/552/553 62–67, 69, 76
SG 751 SAPR 65–70, 97
SG-43 Goryunov machine gun 122
Shepherd scope 154
Short-stroke gas piston 20–21, 25, 28, 45, 48–49, 54, 65, 84, 87–88, 93, 124, 162
SIG 551-A1 65
SIG 716 65

Sioux 107
Six Day War 24, 26, 29, 71–72
"Small Arms Ideal Caliber Panel" 5
SMAW 144
SMLE /Lee-Enfield 110–11, 113, 121, 143–44
Smokeless powder 35, 102, 104, 108, 112, 119
SOCOM 31, 53, 79
SOCOM/SOCOM II (Springfield Armory) 53
Somalia 141
South Africa 17, 23, 68
South America 20, 33, 61, 62
Special Air Service (SAS) 86
Spencer .56 carbine 105–06
Spitzer bullet 102, 108, 149
Springfield Armory (U.S. arsenal) 53
Squad Automatic Weapon (SAW) 19–20, 52, 73, 144, 158, 161
Squad Designated Marksman 168
Squad Scout 53
SR3 (MA) 149–50
SR3M 149
SR-99 75
SS 117
SS109 12, 14, 139, 166
SSK industries 160
Stalin, Joseph 116–17, 118, 121
Stevens, R. Blake 3
StG 44 1, 5–6, 16, 58, 114, 116–18, 124–26, 164
StG 58 30
StGw 57 (SIG) 61
Stick powder 11
Stoner, Eugene 8, 72
Stoner model 63 71–72
Suomi submachine gun 125
Suppressive fire 2, 19, 131
SVD Dragunov 14, 142–44
Switzerland 43, 68

T20 54
T25 54
T44/T44E4 6, 18, 23
T48 18
T65 6, 18
Taiwan 33
Taliban 142–45
Tavor (IWI) 72, 97
Taylor, Maxwell, Gen. (U.S. Army) 8

Tokarev 58, 113–14, 117–19, 121–22, 124–25, 130
Trijicon 154
"Tropical" forend 41, 46
TRW 53
Turkey 43, 47, 97
Type 1/2 receiver 31
Type 57 rifle 53

U.S. Air Force 8
U.S. Army 7–8, 57, 87, 102, 106, 113, 133, 141, 145, 156, 166–67
U.S. intelligence community 10, 150
U.S. Marine Corps 11, 28, 55, 127, 138, 167–68
Urban combat 3, 27, 74, 80, 97, 135, 140, 142, 149–50, 152–53, 167
UZI (IWI) submachine gun 26–27, 71

Venezuela 17, 33
Vickers machine gun 19, 130
Vietnam War 2, 8–9, 12–14, 20, 22, 54–55, 112, 132–33, 139, 141
Volcanic pistol 104
Von Dreyse 109
Vorgrimmler, Ludwig 6
VSK-94 149–50
VSS 149–50

Washington, D.C. 8
Weaver scope 135
Wehrmacht 119, 122
Weimar Republic 114
Western Front 114–15, 120
Westmoreland, William, Gen. (U.S. Army) 10
Wheel lock 101
White House 10
"Whiz kids" 9–11
Winchester 6, 53, 102, 104–09, 111, 133
Winter War (1939–40) 113, 123, 125
World War I 1–2, 19, 103, 110, 112, 114–16, 119, 120, 126, 130
World War II 1–3, 5, 16, 24, 51, 54, 58, 61, 71, 111, 113–15, 119, 126, 130–31, 133, 135
Wyman, Willard, Gen. (U.S. Army) 8

Yount, Harold, Col. (U.S. Army) 10

Zeiss 41

www.ingramcontent.com/pod-product-compliance
Ingram Content Group UK Ltd.
Pitfield, Milton Keynes, MK11 3LW, UK
UKHW042013140426
5217IPUK00015B/1154